Multi-Mode Resonant Antennas

Multi-Mode Resonant
Antennas
Theory, Design, and Applications

Wen-Jun Lu
Lei Zhu

CRC Press
Taylor & Francis Group
Boca Raton London New York

CRC Press is an imprint of the
Taylor & Francis Group, an **informa** business

MATLAB® is a trademark of The MathWorks, Inc. and is used with permission. The MathWorks does not warrant the accuracy of the text or exercises in this book. This book's use or discussion of MATLAB® software or related products does not constitute endorsement or sponsorship by The MathWorks of a particular pedagogical approach to or a particular use of the MATLAB® software.

First edition published 2022
by CRC Press
6000 Broken Sound Parkway NW, Suite 300, Boca Raton, FL 33487-2742

and by CRC Press
4 Park Square, Milton Park, Abingdon, Oxon, OX14 4RN

CRC Press is an imprint of Taylor & Francis Group, LLC

© 2022 Wen-Jun Lu, Lei Zhu

Library of Congress Cataloging-in-Publication Data
Names: Lu, Wen-Jun, 1978- author. | Zhu, Lei, 1963- author.
Title: Multi-mode resonant antennas : theory, design, and applications / Wen-Jun Lu and Lei Zhu.
Description: First edition. | Boca Raton : CRC Press, 2022. |
Includes bibliographical references and index. |
Summary: "This title provides a unique theoretical framework for multi-mode resonant antennas and different approaches to their implementation, with an emphasis on mode gauge functionality, a new concept for a clear identification and flexible control of all usable resonant modes in multi-mode resonant antenna design. The book commences by advancing a generalized odd-even mode theory as a general theoretical framework for resonant elementary antennas, offering new insights into the classical problem of coupling effects between antenna and transmission lines and helping reveal the operation mechanism of elementary antennas under multi-mode resonance. Then, the concept of "mode gauge" is developed and employed for wideband elementary antenna design by simultaneously exciting and tuning multiple resonant modes within a single radiator. Apart from theoretical explorations, the authors also provide analysis of up-to-date implementation of multi-mode resonant elementary antennas with different functionalities, including wideband antennas, circularly polarized antennas, multiband antennas, frequency scanning antennas, and low-profile antennas. Academics, students and professional engineers at all levels will greatly benefit from the book and will be provided with historical background, state-of-the-art methodology, useful design tools, and multiple applications of multi-mode resonant antennas"—Provided by publisher.
Identifiers: LCCN 2021059251 (print) | LCCN 2021059252 (ebook) |
ISBN 9781032271613 (hardback) | ISBN 9781032271637 (paperback) | ISBN 9781003291633 (ebook)
Subjects: LCSH: Antennas (Electronics)
Classification: LCC TK7871.6 .L79 2022 (print) | LCC TK7871.6 (ebook) | DDC 621.382/4—dc23/eng/20220124
LC record available at https://lccn.loc.gov/2021059251
LC ebook record available at https://lccn.loc.gov/2021059252

ISBN: 978-1-032-27161-3 (hbk)
ISBN: 978-1-032-27163-7 (pbk)
ISBN: 978-1-003-29163-3 (ebk)

DOI: 10.1201/9781003291633

Typeset in Minion
by codeMantra

Contents

v

Preface

The discussion of coupled transmission lines and antennas has been nonmathematical because a complete analytical treatment of the difficult coupling problems is unavailable. It has, nevertheless, been based on general electromagnetic principles in so far as a qualitative application of these was possible.

"Coupled Antennas and Transmission
Lines," by R. King, July 1943

Since Ronold King concluded thus in his research article "Coupled Antennas and Transmission Lines" published in *Proceedings of the IRE*, July 1943, over three-quarters of a century has passed, and the analytical treatment of the coupling effect has yet to be solved. The main focus of this book is to advance a *generalized odd-even mode theory* to mathematically model the coupled effect between antennas and feed networks, and physically understand the radiation behavior of antennas under multiple-mode resonance. In this manner, the concept of "one radiator, multiple resonant modes" is presented and employed to design antennas under multiple-mode resonance. Design approaches to basic elementary antennas, i.e., electric dipoles, slot antennas, loop antennas, complementary dipole antennas, and microstrip patch antennas, are respectively developed in a multi-mode resonant sense. Relevant applications of the multi-mode resonant antennas are then raised and discussed. Finally, the advanced multi-mode resonant antenna design approach is compared to other recently developed, popular antenna design approaches in a comprehensive manner.

Chapter 1 advances the generalized odd-even mode theory. It provides a general theoretical framework of the multi-mode resonant antenna design approach. The *interior Green's functions* of the antenna and the source are expanded in terms of eigenfunctions to illustrate the mode-matching (balanced) and mode-mismatching (unbalanced) status in antenna systems. The proposed theory has been well validated by the classical electric dipoles, loop antennas, slot antennas, and microstrip patch antennas. It lays down a unified theoretical framework for all resonant antennas.

Chapter 2 deals with the design approach to multi-mode resonant electric dipole antennas. A brief history of classical resonant electric dipoles is introduced first. Then, the 1-D resonant straight linear dipole and 2-D resonant sectorial dipole are presented, designed, and discussed, respectively. The dual-mode resonant, *full-wavelength sectorial dipole* antenna has been verified to be a high-performance single- or dual-polarized, wide beamwidth antenna. Multi-band designs and recent developments in multi-mode resonant electric dipoles are also introduced with practical examples.

The theory and design approach to multi-mode resonant slot antennas and loop antennas is developed in Chapter 3. A brief history and recent developments of slot antennas and loop antennas are introduced. Dual- and triple-mode resonant slot antennas at their basic forms are presented to validate the multi-mode resonant design approach. Then, the mode transition between slot and loop is discussed. Eigenmode theory of circular loop antenna is developed to illustrate the radiation behavior of loop antennas under a pair of even-mode resonance. In this way, the concept of *Poincaré sphere source antenna* is advanced to depict the bidirectional radiation behavior of circularly polarized loop antenna under a pair of even-mode resonance. Wideband and dual-band even-mode resonant circularly polarized square loop antennas are designed, implemented, and validated.

Chapter 4 explores the complementary dipole antennas under multiple-mode resonances. A brief history of complementary dipole antennas is discussed first. Then, as validated by the loop-dipole combined and the slot-dipole combined antennas, the multi-mode resonant concept can lead to wideband complementary dipole antenna designs with the simplest configuration. A generalized, multi-mode resonant complementary dipole antenna is advanced by transforming the unbalanced electric current flowing on the surface of a planar magnetic dipole antenna. Thus it yields *planar self-balanced magnetic dipole antennas* under both electric and magnetic current resonances.

Chapter 5 begins with the history and development of microstrip patch antennas (MPAs). Then, a *mode gauged design approach* to multi-mode resonant MPAs is developed. By properly adjusting the length and boundary condition of the gauged magnetic dipole, the order of usable resonant modes of circular sector MPAs can be finely tuned and distinctive radiation performance can be realized. The advanced mode gauged design approach has been employed to design compact wideband MPAs, high gain MPAs, tilted beam circularly polarized MPAs, and wideband null frequency scanning MPAs under dual-mode or triple-mode resonances, with a set of closed-form design formulas available for initial key parameters estimation.

Chapter 6 contains design examples of multi-mode resonant antennas for miscellaneous applications, including land mobile/wireless communications, wearable or implantable communications, vehicular communications, millimeter-wave communications, broadcasting, radio frequency identification (RFID), passive coherent location (PCL), etc. As will be validated, the multi-mode resonant antenna design approach still exhibits effectiveness, robustness, and generality even when the antennas' size and their working environments are critically limited.

Chapter 7 presents a summary of the multi-mode resonant antenna design approach, as well as a comprehensive comparison to other popular wideband antenna design approaches that have been developed since the 2000s, in terms of operation principle, brief closed-form design formulas for initial key parameters estimation, external accessory, mode gauged functionality, and configuration complexity. As will be seen, the advanced approach should be novel and distinctive from the existing counterparts.

The design examples in this book have been used in teaching and scientific research at the Nanjing University of Posts and Telecommunications for over one decade. There does not seem to be any miraculous way to eliminate typographical errors or even errors in theory and formulation. The authors hope that there are not too many and will appreciate readers bringing errors that they discover to our attention via the e-mail: wjlu@njupt.edu.cn, so that they can be corrected in any future possible versions.

Acknowledgments

The authors are indebted to Prof. Ya-Ming Bo, Prof. Wei Cao, and Prof. Ming Zhang at the Nanjing University of Posts and Telecommunications for their beneficial discussions and suggestions on this book. This work was supported in part by the National Natural Science Foundation of China under grant no. 61871233 and National Key Research and Development Program of China under grant no. 2021YFE0205900.

Wen-Jun Lu thanks all of his family members, relatives, and friends for their support, love, and encouragement. He expresses deep appreciation to his supervisors, Prof. Hong-Bo Zhu and Prof. Chong-Hu Cheng at the Nanjing University of Posts and Telecommunications, for their guidance, advice, and substantial encouragement over the past two decades. He also thanks his supervised undergraduate and postgraduate students since 2007, for their contributions to the research and development of multi-mode resonant antennas. Special thanks to Ms. Xiao-Hui Mao, Ms. Fei-Yan Ji, Ms. Xiu-Qiong Xing, Ms. Ming-Ge Pan, Ms. Meng-Li Zhao, Ms. Lu Liu, Mr. Jian Yu, Mr. Da-Shuai Zhang, Mr. Xiao Jiang, and Mr. Yong Xiao, his PhD and Master's students, who helped him edit many of the figures and illustrations in this book. Last but not least, special acknowledgment goes to his wife Dr. Xiao-Hui Li, who takes care of their son, provides encouragement and willingness to forgo many other activities so that he could successfully complete the book.

Prof. Zhu expresses his deep appreciation to his former research staff and PhD students at the University of Macau for their involvement and contribution in innovative development of multi-mode antenna concept and technique.

Authors

Wen-Jun Lu is Full Professor at the Nanjing University of Posts and Telecommunications, China. His research interests include antenna theory, antenna designs, and wireless propagation. He is the inventor of generalized odd-even mode theory and planar endfire circularly polarized antennas.

Lei Zhu is Distinguished Professor at the University of Macau, China. His research interests include antenna theory, microwave engineering, and computational electromagnetics. He is the inventor of the multi-mode resonator for wideband circuits/antennas and the numerical open-short-calibration technique.

CHAPTER 1

Generalized Theoretical Framework for Multi-Mode Resonant Antennas

1.1 GENERAL DESIGN GUIDELINES OF RESONANT ANTENNAS: MATHEMATICAL AND PHYSICAL MODELS

In the classical antenna theory, an arbitrary resonant antenna (e.g., electric dipole, slot, loop, microstrip patch antenna, etc.) under investigation should use its principal resonant mode for radiation only, with an external excitation applied at $\bar{r} = \bar{r}'$ that can be mathematically emulated by the Dirac delta function $\delta(\bar{r} - \bar{r}')$. Physically, the Dirac delta function should be expanded into the Fourier series and matched to the antenna on its surface in terms of its resonant modes, i.e., the eigenfunctions (Zhang 1982, Collin 1991), which implies that the antenna system with external feed network should operate under multi-mode resonance. Therefore, a multi-mode resonant antenna design approach based upon the "one radiator, multiple resonant modes" idea and "multi-mode matching" concept should be rigorously advanced and depicted by a general mathematical model. Once such model is readily available, practical examples of dipole,

DOI: 10.1201/9781003291633-1

slot, loop, and microstrip patch antennas (MPAs) will be employed to validate its correctness, effectiveness, and generality.

The problem of simultaneously exciting multiple resonant modes within a single resonator or an antenna can be generally described in the perspective of eigenvalue equation and the *interior Green's function* of $G(\bar{r},\bar{r}')$, which is corresponding to the antenna's surface field/current distribution. In this regard, the "interior Green's function" should be distinguished from the "Green's functions" in free space, open space, or half-open space that have been widely discussed in traditional electromagnetic radiation and scattering problems. The "multi-mode resonance problem" discussed herein should be quite similar to the classical guided wave problem (Rayleigh 1897, Barrow 1936) by solving homogenous or inhomogeneous wave equations under closed interval, finite closed cross section or closed cavity with specific boundary conditions. As extensively formulated in many classical textbooks, this is a boundary value problem of Helmholtz's equation under specific finite-range, closed boundary conditions (Zhang 1982, Collin 1991).

Generally, suppose that the resonator or antenna under investigation should have an arbitrary size with a source of excitation at $\bar{r}=\bar{r}'$, thus the interior problem can be mathematically defined on a closed, finite interval of V with homogeneous boundary conditions on its smooth, twice continuous differentiable bounded surface ∂V, such that

$$-\left(\nabla^2+k^2\right)G(\bar{r},\bar{r}')=\delta\left(\bar{r}-\bar{r}'\right),\bar{r},\bar{r}'\in V$$

$$\left.\alpha\frac{\partial G}{\partial n_s}+\beta G=0\right|_{\bar{r}\in\partial V}$$

(1.1a)

where n_s denotes the direction of the outer normal vector of the bounded surface ∂V. Correspondingly, the homogeneous equation and boundary condition of the resonator/antenna's eigenmodes should satisfy

$$-\left(\nabla^2+k_n^2\right)\psi(\bar{r})=0,\bar{r},\bar{r}'\in V$$

$$\left.\alpha\frac{\partial\psi}{\partial n_s}+\beta\psi=0\right|_{\bar{r}\in\partial V}$$

(1.1b)

where $\left\{\psi_n(\bar{r})\right\}$ is the full, discrete set of accordingly defined eigenfunctions, corresponding to the discrete eigenvalue (Square of characteristic wave number) set of $\left\{k_n^2\right\}$.

In fact, the resonator or antenna can be excited by different kinds of sources. Besides the conventional unit Dirac delta function $\delta(\bar{r}-\bar{r}')$ (i.e., unit impulse), the unit doublet function (i.e., the first-order derivative of Dirac delta function) $\delta'(\bar{r}-\bar{r}')$ can also serve as a source of excitation. Suppose the source of excitation locates at $\bar{r}=\bar{r}'$ and use the general expansion expression in terms of eigenfunctions (Zhang 1982, Collin 1991), as well as the sampling property of the $\delta'(\bar{r}-\bar{r}')$ function, i.e., $\iiint_V \delta'(\bar{r}-\bar{r}')f(\bar{r})=-f'(\bar{r}')$, the surface current/field distributions under unit impulse/doublet functions excitation can be express by Eqs. (1.2a) and (1.2b), respectively.

$$G_{\text{impulse}}(\bar{r},\bar{r}')=\sum_{n}\frac{\psi_n(\bar{r}')\psi_n(\bar{r})}{\left(k_n^2-k^2\right)\|\psi_n\|^2} \tag{1.2a}$$

$$G_{\text{doublet}}(\bar{r},\bar{r}')=\sum_{n}\frac{-\psi_n'(\bar{r}')\psi_n(\bar{r})}{\left(k_n^2-k^2\right)\|\psi_n\|^2} \tag{1.2b}$$

$$G(\bar{r},\bar{r}')=aG_{\text{impulse}}(\bar{r},\bar{r}')+bG_{\text{doublet}}(\bar{r},\bar{r}')$$

$$=\sum_{n}\frac{\left[a\psi_n(\bar{r}')-b\psi_n'(\bar{r}')\right]\psi_n(\bar{r})}{\left(k_n^2-k^2\right)\|\psi_n\|^2} \tag{1.2c}$$

Physically, the even-symmetric, unit impulse function $\delta(\bar{r}-\bar{r}')$ represents the "differential mode" excitation, while the odd-symmetric, unit doublet function $\delta'(\bar{r}-\bar{r}')$ represents the "common mode" one, as shown in Figure 1.1a. Arbitrary excitations can be represented by the linear combination of $\delta(\bar{r}-\bar{r}')$ and $\delta'(\bar{r}-\bar{r}')$ functions. Thus the surface current/field distribution of an arbitrary antenna under arbitrary source excitation of $a\delta(\bar{r}-\bar{r}')+b\delta'(\bar{r}-\bar{r}')$ (where a, b are arbitrary constants) at $\bar{r}=\bar{r}'$ can be generally represented by the interior Green's function in Eq. (1.2c).

In fact, since most elementary antennas exhibit symmetry at least in one dimension, it is without loss of generality that the generalized odd-even mode theory can be derived in a clearly understandable fashion by considering arbitrarily shaped, symmetric dipoles with center feed. Suppose that the feed point $\bar{r}=\bar{r}'$ should locate at the symmetrical center of the antenna. As shown in the cases of Figure 1.1b, c and d, e, the two terminals of the antenna are driven in differential/common E/M-mode, respectively. Since the impulse function is even-symmetric, while the

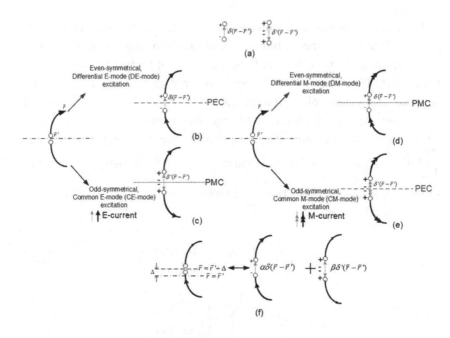

FIGURE 1.1 Generalized odd-even mode theory for arbitrary antennas under multi-mode resonance. (a) Sources with different symmetry. (b) Differential E-mode. (c) Common E-mode. (d) Differential M-mode. (e) Common M-mode. (f) Off-center fed symmetric dipole.

doublet one is odd-symmetric, they can only excite the even-symmetric and odd-symmetric eigenmodes of the antenna, respectively. In this manner, it is reasonable to elaborate the generalized odd-even mode theory in the following aspects:

i. The interior Green's function G can be recognized as the combination of two parts, i.e., G_e and G_o, which correspond to the even-symmetric and odd-symmetric parts, respectively. The even-symmetric part G_e can be differentially excited by an even-symmetric source function (e.g., $\delta(\bar{r}-\bar{r}')$) only, which corresponds to the case of $a=1$ and $b=0$ in Eq. (1.2c); while the odd-symmetric part G_o can be commonly excited by an odd-symmetric source function (e.g., $\delta'(\bar{r}-\bar{r}')$) only, which corresponds to the case of $a=0$ and $b=1$ in Eq. (1.2c).

ii. Accordingly, the eigenfunction set of $\{\psi_n(\bar{r})\}$ can be classified into two subsets of $\{\psi_{ne}(\bar{r})\}$ and $\{\psi_{no}(\bar{r})\}$, as well as the eigenvalue set of

$\{k_n{}^2\}$ can be classified into $\{k_{ne}{}^2\}$ and $\{k_{no}{}^2\}$. They would be employed to expand G_e and G_o, respectively.

iii. Using the general spectral expansion expression in terms of eigenfunctions (Zhang 1982, Collin 1991) and Eq. (1.2c), G_e and G_o can be accordingly expanded as

$$G_e(\vec{r},\vec{r}') = \sum_n \frac{\psi_{ne}(\vec{r}')\psi_{ne}(\vec{r})}{(k_{ne}^2 - k^2)\|\psi_{ne}\|^2} \tag{1.3a}$$

$$G_o(\vec{r},\vec{r}') = \sum_n \frac{-\psi'_{no}(\vec{r}')\psi_{no}(\vec{r})}{(k_{no}^2 - k^2)\|\psi_{no}\|^2} \tag{1.3b}$$

As can be seen, the even-symmetric interior Green's function in Eq. (1.3a) is fully identical to the general expression described in classical textbooks (Zhang 1982, Collin 1991). Equation (1.3b) exactly describes the odd-symmetric interior Green's function excited by a unit doublet electric current sheet. Note that Eqs. (1.3a) and (1.3b) are 1-D spectral representations obtained by the generalized odd-even mode theory illustrated in Figure 1.1a–e: They can be elaborated into 2-D and 3-D cases under different orthogonal curvilinear coordinate systems, respectively. In practical applications, the interior Green's function defined on closed finite intervals can be generalized and represented by multiple infinity series, and employed to deduce the surface current/E-field distributions accordingly. Once the surface source distribution is readily known, the vector and scalar potential functions, as well as the electric and magnetic fields in the near- and far-field zones (i.e., "the external problem"), can be correspondingly formulated (Harrington 1961).

For an arbitrary-shaped, symmetrically fed electric dipole, when it is under even-symmetric, pure differential E-mode (DE-mode) excitation, its plane of symmetry should exhibit a natural boundary condition (Lu and Zhu 2015a) of perfect electric conductor or electric wall, as shown in Figure 1.1b. Under odd-symmetric, pure common E-mode (CE-mode) excitation, the plane of symmetry of a symmetric dipole antenna should exhibit a natural boundary condition of perfect magnetic conductor or magnetic wall, as shown in Figure 1.1c. In the symmetrically fed case, unit impulse electric current sheet $\delta(\vec{r}-\vec{r}')$ would never excite CE-modes, while unit doublet electric current sheet $\delta'(\vec{r}-\vec{r}')$ would never excite DE-modes.

The previous analyses on multi-mode resonant electric dipoles are also valid for the cases of magnetic dipoles, which can be easily attained according to the duality of electric and magnetic fields, as shown in Figure 1.1d and e. It is seen the cases of magnetic dipoles are complementary to the electric ones: Under even-symmetric, pure differential M-mode (DM-mode) excitation, the plane of symmetry of a symmetric magnetic dipole antenna should exhibit a natural boundary condition (Lu and Zhu 2015a) of magnetic wall, as shown in Figure 1.1d. Under odd-symmetric, pure common M-mode (CM-mode) excitation, the plane of symmetry of a symmetric magnetic dipole antenna should exhibit a natural boundary condition of electric wall, as shown in Figure 1.1e. Note that the source functions in Figure 1.1d and e should be virtual magnetic current sheets of unit strength. According to the principle of duality and the principle of equivalence (Harrington 1961), the virtual magnetic sources could be respectively supplanted by the orthogonal, physically realizable electric current sheets at their exact locations, according to Ampère's right-hand grip rule. In the symmetrically fed case, unit impulse magnetic current sheet $\delta(\vec{r}-\vec{r}')$ would never excite CM-modes, while unit doublet magnetic current sheet $\delta'(\vec{r}-\vec{r}')$ would never excite DM-modes.

As is seen, the terms of "differential mode" and "common mode" widely used in conventional odd-even mode theory have been respectively extended and elaborated into the "differential E-mode", "common E-mode", "differential M-mode", and "common M-mode" illustrated in Figure 1.1. In theory, the operation principle as well as the working behavior of arbitrary antennas under arbitrary excitations can be intuitively illustrated by the combination of even- and odd-symmetric eigenmodes.

$$G_e(\vec{r},\vec{r}'+\Delta)=G_{ee'}(\vec{r},\vec{r}'+\Delta)+G_{eo'}(\vec{r},\vec{r}'+\Delta)$$

$$=\sum_n \frac{\psi_{ne}(\vec{r}'+\Delta)\psi_{ne}(\vec{r})}{(k_{ne}^2-k^2)\|\psi_{ne}\|^2}-\sum_n \frac{\psi'_{ne}(\vec{r}'+\Delta)\psi_{ne}(\vec{r})}{(k_{ne}^2-k^2)\|\psi_{ne}\|^2} \quad (1.4a)$$

$$G_o(\vec{r},\vec{r}'+\Delta)=G_{oe'}(\vec{r},\vec{r}'+\Delta)+G_{oo'}(\vec{r},\vec{r}'+\Delta)$$

$$=\sum_n \frac{\psi_{no}(\vec{r}'+\Delta)\psi_{no}(\vec{r})}{(k_{no}^2-k^2)\|\psi_{no}\|^2}-\sum_n \frac{\psi'_{no}(\vec{r}'+\Delta)\psi_{no}(\vec{r})}{(k_{no}^2-k^2)\|\psi_{no}\|^2} \quad (1.4b)$$

Let's consider the final case, i.e., the symmetric dipole with off-center feeding at $\bar{r} = \bar{r}' + \Delta$ shown in Figure 1.1f. Since the feed point does not coincide with the symmetric center, the impulse function can partially excite the odd-symmetric modes, as well as the doublet function can partially excite the even-symmetric ones in the same manner. Therefore, the interior Green's function can be described by the combinations of four parts as shown in Eqs. (1.4a) and (1.4b): $G_{ee'}$, $G_{eo'}$, $G_{oe'}$, and $G_{oo'}$. The prime in the subscript denotes the symmetry/parity of the source. When the off-center distance $\Delta = 0$ (i.e., center feeding), $G_{eo'}$ and $G_{oe'}$ in Eqs. (1.4a) and (1.4b) would vanish, thus $G_{ee'}$ and $G_{oo'}$ would degenerate into the special cases in Eqs. (1.3a) and (1.3b). As can be seen, in the off-center fed case, the even-symmetric source can partially excite the odd-symmetric eigenmodes that can be represented by $G_{oe'}$, as well as the odd-symmetric source can partially excite the even-symmetric ones that can be represented by $G_{eo'}$.

As shown in Figure 1.1f, let's consider the ratio of DE-modes and CE-modes under different excitations further. Suppose only a unit impulse function is applied at the off-center position of Δ, thus the interior Green's function can be represented by $G_{ee'}$, G_o and $G_{oe'}$

$$G\left(\bar{r},\bar{r}'+\Delta\right)=G_{ee'}\left(\bar{r},\bar{r}'+\Delta\right)+G_{oe'}\left(\bar{r},\bar{r}'+\Delta\right)$$

$$=\sum_n \frac{\psi_{ne}\left(\bar{r}'+\Delta\right)\psi_{ne}\left(\bar{r}\right)}{\left(k_{ne}^2-k^2\right)\left\|\psi_{ne}\right\|^2}+\sum_n \frac{\psi_{no}\left(\bar{r}'+\Delta\right)\psi_{no}\left(\bar{r}\right)}{\left(k_{no}^2-k^2\right)\left\|\psi_{no}\right\|^2} \quad (1.5a)$$

$$G\left(\bar{r},\bar{r}'\right)=\alpha G_e\left(\bar{r},\bar{r}'\right)+\beta G_o\left(\bar{r},\bar{r}'\right)$$

$$=\sum_n \frac{\alpha\psi_{ne}\left(\bar{r}'\right)\psi_{ne}\left(\bar{r}\right)}{\left(k_{ne}^2-k^2\right)\left\|\psi_{ne}\right\|^2}-\sum_n \frac{\beta\psi_{no}'\left(\bar{r}'\right)\psi_{no}\left(\bar{r}\right)}{\left(k_{no}^2-k^2\right)\left\|\psi_{no}\right\|^2} \quad (1.5b)$$

$$u_e\left(k,\Delta\right)=\left|\frac{\beta}{\alpha}\right|=\left|\frac{G_e\left(\bar{r},\bar{r}'\right)G_{oe'}\left(\bar{r},\bar{r}'+\Delta\right)}{G_e\left(\bar{r},\bar{r}'+\Delta\right)G_o\left(\bar{r},\bar{r}'\right)}\right|$$

$$=\left|\frac{\displaystyle\sum_n \frac{\psi_{ne}\left(\bar{r}'\right)\psi_{ne}\left(\bar{r}\right)}{\left(k_{ne}^2-k^2\right)\left\|\psi_{ne}\right\|^2}\sum_n \frac{\psi_{no}\left(\bar{r}'+\Delta\right)\psi_{no}\left(\bar{r}\right)}{\left(k_{no}^2-k^2\right)\left\|\psi_{no}\right\|^2}}{\displaystyle\sum_n \frac{\psi_{ne}\left(\bar{r}'+\Delta\right)\psi_{ne}\left(\bar{r}\right)}{\left(k_{ne}^2-k^2\right)\left\|\psi_{ne}\right\|^2}\sum_n \frac{\psi_{no}'\left(\bar{r}'\right)\psi_{no}\left(\bar{r}\right)}{\left(k_{no}^2-k^2\right)\left\|\psi_{no}\right\|^2}}\right| \quad (1.5c)$$

$$u_o(k,\Delta) = \left|\frac{\beta}{\alpha}\right| = \left|\frac{G_e(\vec{r},\vec{r}')G_o(\vec{r},\vec{r}'+\Delta)}{G_{eo'}(\vec{r},\vec{r}'+\Delta)G_o(\vec{r},\vec{r}')}\right|$$

$$= \left|\frac{\displaystyle\sum_n \frac{\psi_{ne}(\vec{r}')\psi_{ne}(\vec{r})}{(k_{ne}^2 - k^2)\|\psi_{ne}\|^2} \sum_n \frac{\psi_{no}(\vec{r}'+\Delta)\psi_{no}(\vec{r})}{(k_{no}^2 - k^2)\|\psi_{no}\|^2}}{\displaystyle\sum_n \frac{\psi_{ne}'(\vec{r}'+\Delta)\psi_{ne}(\vec{r})}{(k_{ne}^2 - k^2)\|\psi_{ne}\|^2} \sum_n \frac{\psi_{no}'(\vec{r}')\psi_{no}(\vec{r})}{(k_{no}^2 - k^2)\|\psi_{no}\|^2}}\right| \qquad (1.5d)$$

Equivalently, the effect of off-center feeding can be expressed by incorporating a hybrid source excitation $\alpha\delta(\vec{r}-\vec{r}')+\beta\delta'(\vec{r}-\vec{r}')$ at the center of $\vec{r}=\vec{r}'$, according to Eq. (1.2c) and Figure 1.1f. Using Eqs. (1.3a) and (1.3b), the interior Green's function can be alternatively represented by Eq. (1.5b). Thus, the coefficients of α and β can be determined by comparing the corresponding terms in Eqs. (1.5a) and (1.5b). Therefore, the ratio of CE-modes and DE-modes under off-center, even-symmetric excitation can be described by introducing a function of $u_e(k, \Delta)$, as shown in Eq. (1.5c). In the same way, if a unit doublet source is applied at the acentric position of Δ in Figure 1.1f instead, the ratio of CE-modes and DE-modes under acentric, odd-symmetric excitation can also be described in terms of G_e, G_o and $G_{eo'}$ by introducing the function $u_o(k, \Delta)$, as shown in Eq. (1.5d).

According to Eq. (1.5), it can be concluded that the "mutual coupling" or "cross coupling" terms of $G_{eo'}$ and $G_{oe'}$ in the interior Green's function can be employed to elaborate the degree of asymmetry/unbalance (denoted as u_e and u_o, respectively) of a symmetric dipole under off-center excitation: *In ultimate, the unbalanced component of an antenna system should be mathematically dominated by the cross-coupled term in the interior Green's function.* In the opinion of antenna designs, the generalized odd-even mode theory bridges the antenna in tandem with the feed network (i.e., the source), and it paves a promising way to mathematically elaborate the mutual coupling effect between them (King 1943).

Let's further compare the generalized odd-even mode theory to the classical odd-even mode theory. As is well known, the classical odd-even mode analysis approach in microwave circuit theory is formulated based upon the transmission line theory. It has an important prerequisite postulate that both circuit (i.e., resonator, antenna) and source (i.e., feed network, transmission line) should be perfectly mode-matched (balanced) under single-mode excitation. In the generalized odd-even mode theory, such prerequisite postulate is no longer required: Both mode-matched (balanced) and mode-mismatched (unbalanced) cases between circuit (resonator, antenna) and source (feed network, transmission line) can be described in a more general sense by using Eqs. (1.3)–(1.5). In this way, the behavior of circuits (resonator, antenna) under multiple resonant modes excitation, or the coupling effect between antenna and feeders, can be depicted in a more general perspective of electromagnetic fields and spectral expansion theory of interior Green's function in terms of eigenmode functions: In the balanced case, classical interior Green's function represented by Eq. (1.2a) can be employed. In the unbalanced case, elaborated interior Green's functions with cross-coupled terms represented by Eqs. (1.2b) and (1.3)–(1.5) would be required.

1.2 TYPICAL MULTI-MODE RESONANT ANTENNA EXAMPLES

In the sense of engineering applications and without loss of generality, we will use the simplest, but as well as the most important and prevailed elementary antennas, to validate the generalized odd-even mode theory in the following sections. Since more complicated antennas can be treated as the combinations of the simplest elementary antennas, for brevity, they would not be discussed in detail.

1.2.1 Straight Resonant Electric Dipole with Both Ends Open/Short-Circuited

The simplest, ideal center-fed electric dipole shown in Figure 1.2 is considered as the first example. The integro-differential equation and the χ-theory pioneered by (Leontovich and Levin 1944) are paralleled to the Ω-theory (Schelkunoff 1945, King 2002) pioneered by Hallén in the same era. Suppose the dipole is linearly slim (i.e., with its radius a and half-length L satisfies $a \ll L$), the precise current distribution $J(z)$ can be solved by employing the method of perturbation. In this manner, a power series of χ-factor (defined as $\chi = 1/[2\ln(ka)]$) can be utilized to expand $J(z)$, where a is the radius of the dipole, k is the wave number, and $J_0(z)$, $J_1(z)$, and $J_2(z)$ indicate the eigenmode current distribution without feed line, current distributions that are excited at the antinode and node, respectively.

$$J(z) = J_0(z) + \chi J_1(z) + \chi^2 J_2(z) + \cdots \qquad (1.6)$$

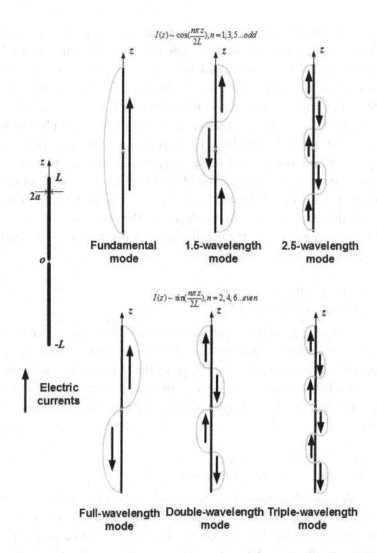

FIGURE 1.2   The previous six resonant modes of a symmetric electric dipole antenna.

Here, only the resonant component of $J_0(z)$ is considered. Using the boundary condition of the dipole antenna at both ends, i.e., $J(\pm L)=0$, the homogeneous Helmholtz's equation for the eigenmodes of resonant currents (regardless of the feed line) J_0 can be written as

$$J_0''(z)+k_n^2 J_0(z)=0, J_0(\pm L)=0 \qquad (1.7)$$

$$J_0(z) = \begin{cases} J_0 \cos k_n z, n \text{ is odd} \\ J_0 \sin k_n z, \ n \text{ is even} \end{cases}$$

$$k_n = n\frac{\pi}{2L}, n = 1,2,3\ldots \tag{1.8}$$

The eigenmode functions $J_0(z)$ and the associated characteristic wave numbers k_n (real, positive numbers with a dimension of m^{-1}) of resonant current modes represented by Eq. (1.8) can be substituted into Eq. (1.3) and led to precise surface current distributions under different excitations. Thus the behavior of multi-mode resonant dipole under different source excitations can be described. Figure 1.2 clearly demonstrates that the resonant eigenmodes should obey a "cosine-sine alternating discipline" (Lu et al. 2019): All odd-order resonant modes (DE-modes) should exhibit cosine-dependency and even-symmetry, while all even-order ones (CE-modes) should alternatively exhibit sine-dependency and odd-symmetry. When the dipole is centered fed at $z' = 0$ by a current sheet of $\delta(z)$ (corresponding to the case of Figure 1.1a), substituting the cosine functions into Eq. (1.3a), it is obvious that the current distribution should be contributed by all cosine-dependent DE-modes with even-symmetry that can be sufficiently excited with non-zero coefficients. If we substitute the sine functions into Eq. (1.3a), since the coefficients $\sin(n\pi z'/L)|_{z'=0} \equiv 0$, hence all sine-dependent CE-modes with odd-symmetry would never be excited under DE-mode excitation. This well interprets the reason for all odd-symmetric, even-order dipole modes (e.g., full-wavelength mode, double-wavelength mode, etc.) would be fully suppressed in center-fed dipole antennas.

Alternatively, if the dipole is centered fed at $z' = 0$ by a unit doublet function (corresponding to the case of Figure 1.1b), substituting the sine functions into Eq. (1.3b) and since the coefficients of $\cos(n\pi z'/L)|_{z'=0} \equiv 1$, hence all sine-dependent, CE-modes with odd-symmetry would be sufficiently excited under CE-mode excitation. Therefore, a full-wavelength dipole should be effectively driven by a pair of ports with a common terminal, or an asymmetrically off-centered feed (i.e., $z' \neq 0$): Recently, experimental results for dual-port, full-wavelength dipoles (Khraisat et al. 2012) have been presented to validate the effective excitation of full-wavelength mode shown in Figure 1.2.

Let's further consider an acentric-fed, dual-mode resonant dipole with half-wavelength (i.e., $\psi_e = \cos\dfrac{\pi}{2L}z$) and full-wavelength (i.e., $\psi_o = \sin\dfrac{\pi}{L}z$) modes excited in simultaneous. Suppose the dual-mode resonant dipole is excited by a unit impulse source at $z' = \xi(0 \leq \xi \leq L)$. In this case, the complex expression of Eq. (1.5c) can be simplified and approximately expressed as

$$u_e(\xi) = \left|\frac{\beta}{\alpha}\right| \approx \frac{L\sin\dfrac{\pi}{L}\xi}{\pi\cos\dfrac{\pi}{2L}\xi} = \frac{2L}{\pi}\sin\frac{\pi}{2L}\xi, \quad 0 \leq \xi \leq L \qquad (1.9)$$

As can be seen from Eq. (1.9), in the center-fed case ($\xi = 0$), the degree of unbalance u_e should be zero when the antenna is fully excited at DE-modes. The modes of antenna and feed line perfectly match with each other. With the increment of the feed position ξ, the degree of unbalance u_e should gradually increase, and the antenna should operate under both DE- and CE-mode resonance. Especially, for a half-wavelength dipole with $2L = \pi$, u_e should vary from 0 to unity. Therefore, the generalized odd-even mode theory has been successfully employed to illustrate the multi-mode resonant excitation and unbalanced situation of the simplest, straight electric dipole.

If the boundary condition at both ends of the dipole are short-circuited, i.e., $\dfrac{\partial J_0(z)}{\partial z}\bigg|_{z=\pm L} = 0$, the cosine- and sine-alternating discipline would be still valid but interchanged (Schelkunoff 1945)

$$J_0(z) = \begin{cases} J_0 \sin k_n z, n \text{ is odd} \\ J_0 \cos k_n z, n \text{ is even} \end{cases}$$

$$k_n = n\frac{\pi}{2L}, n = 1, 2, 3\dots \qquad (1.10a)$$

$$k_n = \frac{2\pi}{\lambda_n}, \phi \in [-\pi, \pi], 2\pi a = n\frac{\lambda_n}{2}, \frac{\partial J_0(\phi)}{\partial \phi}\bigg|_{\phi=\pm\pi} = 0, n = 1, 2, 3\dots \quad (1.10b)$$

$$J_0(\phi) = J_0\begin{cases} \sin k_n(\phi a) \\ \cos k_n(\phi a) \end{cases} \Rightarrow J_0(\phi) = \begin{cases} J_0 \sin \dfrac{n}{2}\phi, n \text{ is odd} \\ J_0 \cos \dfrac{n}{2}\phi, n \text{ is even} \end{cases} \qquad (1.10c)$$

A resonant electric dipole with both ends short-circuited should be a resonant loop antenna, as indicated in (Schelkunoff and Friis 1952). Thus the case of Eq. (1.10a), with boundary conditions of Eq. (1.10b), can be accordingly modified into Eq. (1.10c) and employed to depict the circumferential current distribution of an ideally lossless, resonant circular loop antenna with radius a. As the second example, both ends of the straight dipole have been bent to coincide at $\phi=\pm\pi$. Suppose the circular loop antenna is fed at $\phi=0$, with its circumference $2\pi a$ must be the integer multiples of the one-half wavelength, Eq. (1.10c) can be attained. The case of Eq. (1.10c) exactly describes the odd, sine-dependent CE-modes and, even, cosine-dependent DE loop modes. Using the eigenfunctions in Eqs. (1.10c) and (1.3b), it is easy to verify that CE-modes with sine current distributions could be commonly excited by a pair of current sheets of $\delta'(\phi)$ near its current node of $\phi=0$. While DE-modes with cosine current distributions can be differentially excited by a current sheet of $\delta(\phi)$ at its current antinode $\phi=0$, as illustrated by Eqs. (1.10c) and (1.3a). Unlike the axial-symmetric straight dipole, the circular loop is a rotationally, centro-symmetric configuration as well. Therefore, further discussions on circular loop antennas will be carried out in Chapter 3.

By carefully comparing the two examples, we can find that the symmetry of eigenmodes of fundamental dipoles is independent of the external excitation, but hinges on the boundary conditions at both ends only. When the symmetry (alternatively, the natural boundary condition) of eigenmodes can be matched to that of a source of excitation (Lu et al. 2019, You et al. 2016), the eigenmodes would be sufficiently excited to resonant, otherwise, they would be suppressed or evanescent instead.

1.2.2 Resonant Magnetic Dipoles: Slot Antenna and Microstrip Patch Antenna

As convinced from the results of centered- and offset-fed slotline antennas (Lu and Zhu 2015a, Wang S. G. et al. 2017, Wang H. et al. 2017), the DM- and CM-modes can be simultaneously excited within a single magnetic dipole by a microstrip feed line with a stub. The cases of DM- and CM-modes are simply complementary to the electric ones, and the slotline antennas previously discussed (Lu and Zhu 2015a, Wang S. G. et al. 2017, Wang H. et al. 2017) can be treated as special cases of those shown in Figure 1.1c and d. In the following section, let's continue to discuss the 2-D case of MPAs.

14 ■ Multi-Mode Resonant Antennas

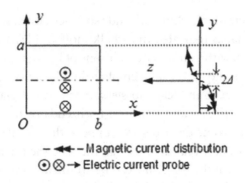

− ◄◄ − Magnetic current distribution
⊙ ⊗ → Electric current probe

FIGURE 1.3 The TM_{01} mode resonant rectangular MPA under single-ended and differential excitations.

Let's consider the rectangular MPA shown in Figure 1.3, with size $b \times a$, and tuned to resonate at its dominant TM_{01} mode. The magnetic current distribution can be expressed by the eigenfunction as

$$\bar{M}_y\left(y\right)=-\bar{y}\frac{E_0}{k^2-\left(\dfrac{\pi}{a}\right)^2}\cos\frac{\pi}{a}y=\bar{y}E_{01}\cos\frac{\pi}{a}y,\ 0<y<a \qquad (1.11)$$

As a common sense, when the MPA is fed by a single electric current probe at the magnetic current antinode $y'=0$ or $y'=a$, TM_{01} mode can be excited in sufficient because its eigenmode function is even-symmetric about xz-plane at $y'=0$ or $y'=a$. When the probe is placed near the node of magnetic current distribution, i.e., $y'=a/2$, TM_{01} mode could hardly be excited, which can be well mathematically explained by using Eq. (1.3a), since $\cos\frac{\pi}{a}y'\big|_{y'=\frac{a}{2}}=0$. As is seen, this is exactly a special DM-mode excitation case of that shown in Figure 1.1c by replacing the magnetic current source with an orthogonal, electric counterpart according to Ampère's right-hand grip rule.

Alternatively, let's consider the case of how the MPA could be effectively excited at its magnetic current node of $y'=a/2$. A pair of orthogonal, z-oriented electric current probes should be introduced near $y'=a/2$: Note that the separation Δ is approaching 0 so that the co-located probes pair could emulate the doublet function of $\delta'\left(y-\dfrac{a}{2}\right)$. With reference to Figure 1.1d, TM_{01} mode can also be excited in theory:

As mathematically calculated by Eq. (1.3b), a non-zero coefficient of

$$-\left(\cos\frac{\pi}{a}y\right)'\bigg|_{y=\frac{a}{2}} = \frac{\pi}{a}\sin\frac{\pi}{a}y\bigg|_{y=\frac{a}{2}} = \frac{\pi}{a}$$ can be attained. Therefore, Figure 1.1d

and Eq. (1.3b) can be employed to well explain the operation principle of differentially driven MPAs (Cheston et al. 1970, Chiba et al. 1982, Zhang 2007): Ideally, a pair of differential electric current probes should be co-located arranged at the magnetic current node of the desired resonant mode. In practice, a pair of differential probes should be symmetrically arranged about the node or nodal line (i.e., $\Delta \neq 0$), for better impedance matching. Indeed, this can be recognized as an alternative CM-mode excitation case of that shown in Figure 1.1d by replacing the magnetic current sources with their orthogonal, electric counterparts.

Besides the differential electric current probes, a TM_{01} mode resonant MPA can be excited at the center by a small coupled aperture (Pozar 1985): The aperture can exhibit a natural boundary condition matched to the one of the desired magnetic current distribution, just as the differential E-current probes behave (You et al. 2016): By comparing these cases, it is found that both aperture-coupled and differential probes can offer a null-plane (i.e., electric wall) boundary condition within the symmetric plane of the patch. Such null-plane boundary condition is perfectly matched to the desired one of resonant TM_{01} mode. It is further confirmed that the boundary conditions of the TM_{01} mode MPAs and the feed networks (i.e., differential E-current probes and coupled aperture) should be well matched with each other (You et al. 2016) within other principal-cut planes. It is easy to verify that when the MPA resonates at high-order modes, the aforementioned analyses based upon eigenfunctions and Eq. (1.3) should be still valid: *The rule of thumb is that to qualitatively match the respective natural boundary conditions of the desired resonant mode and the feed network at the position of excitation.*

Till now, typical elementary antennas (i.e., symmetric dipole, circular loop, slot antennas, MPAs with single-end fed, MPAs with differential probes or aperture-coupled fed) corresponding to all cases illustrated in Figure 1.1 have been studied. As tabulated and comprehensively compared in Table 1.1, the generalized odd-even mode theory can be evidently convinced by correspondingly matching the antennas and excitations to the general cases. If we compare the 1-D straight dipole, circular loop, slot antennas, and 2-D rectangular MPA cases, it can be further found out that

TABLE 1.1   Resonant Antennas Comparisons in Terms of Excitations

Antennas	Excitations	General Case(s)
Center-fed dipole (Leontovich et al. 1943)/Dual-port dipole (Khraisat et al. 2012)	CE-mode/DE-mode	Figure 1.1b and c
Circular loops (Schelkunoff and Friis 1952)	CE-mode/DE-mode	Figure 1.1b and c
Center-fed slots (Lu et al. 2015)	DM-mode	Figure 1.1d
Acentric-fed slots (Wang S. G. et al. 2017, Wang H et al. 2017)	DM-mode and CM-mode	Figure 1.1d and e
Differentially fed MPAs (You et al. 2016, Cheston et al. 1970, Chiba et al. 1982, Zhang 2007)	CM-mode	Figure 1.1e

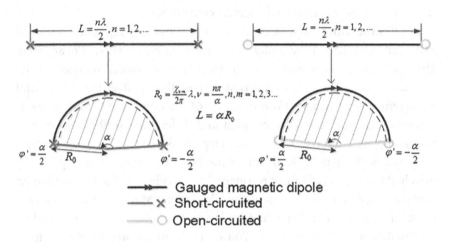

FIGURE 1.4   Mode mapping from rectangular to cylindrical coordinates: 2-D multi-mode resonant magnetic dipoles and mode gauged design approach to circular sector patch antennas.

an even-symmetric, impulse-type source function like $\delta(\bar{r})$ could be introduced to excite the resonant modes at their current/field antinodes, while an odd-symmetric, doublet-type source function like $\delta'(\bar{r})$ could be possibly introduced to excite the resonant modes near their current/field nodes.

As the extensive application of generalized odd-even mode theory as well as further validations, let's revisit the case of multi-mode resonant circular sector patch antennas, which are designed according to the length of a gauged magnetic dipole, and symmetrically fed at the angular bisector (Lu et al. 2017, 2018). The magnetic current distributions of the circular sector patch under $TM_{\nu,\,m}$ modes resonance can be expanded by the 2-D eigenfunctions in cylindrical coordinates, i.e., the Fourier-Bessel series. In the case that both radii short-circuited (Lu et al. 2018), the magnetic current distribution should satisfy

$$
\bar{M}_\phi(\rho,\phi)=\begin{cases}\hat{\phi}\sum_n^{odd}\sum_m\dfrac{E_\nu J_\nu(k\rho)\cos\nu\phi}{k^2-k_{\nu,m}^2}, & \nu=\dfrac{n\pi}{\alpha},n=1,3,5\ldots,m=1,2\ldots\\[4mm]\hat{\phi}\sum_n^{even}\sum_m\dfrac{E_\nu J_\nu(k\rho)\sin\nu\phi}{k^2-k_{\nu,m}^2}, & \nu=\dfrac{n\pi}{\alpha},n=2,4,6\ldots,m=1,2\ldots\end{cases}
$$

$$
0\le\rho\le R_0,-\frac{\alpha}{2}\le\phi\le\frac{\alpha}{2}
$$

(1.12)

In the case that both radii are open-circuited (Lu et al. 2017), the magnetic current distribution should interchangeably satisfy

$$
\bar{M}_\phi(\rho,\phi)=\begin{cases}\hat{\phi}\sum_n^{odd}\sum_m\dfrac{-E_\nu J_\nu(k\rho)\sin\nu\phi}{k^2-k_{\nu,m}^2}, & \nu=\dfrac{n\pi}{\alpha},n=1,3,5\ldots,m=1,2\ldots\\[4mm]\hat{\phi}\sum_n^{even}\sum_m\dfrac{E_\nu J_\nu(k\rho)\cos\nu\phi}{k^2-k_{\nu,m}^2}, & \nu=\dfrac{n\pi}{\alpha},n=2,4,6\ldots,m=1,2\ldots\end{cases}
$$

$$
0\le\rho\le R_0,-\frac{\alpha}{2}\le\phi\le\frac{\alpha}{2}
$$

(1.13)

As is seen, Eqs. (1.12) and (1.13) are indeed the special, 2-D cases of Eqs. (1.3a) and (1.3b) in a cylindrical coordinate system, as illustrated in Figure 1.4. As is described in Eq. (1.12), the odd-order magnetic current modes exhibit cosine-dependent circumferential components. They should be DM-modes and could be excited by a coaxial current probe placed at the angular bisector $\phi=0$. The even-order magnetic current modes exhibit sine-dependent circumferential components, which implies that they should be CM-modes and could hardly be excited by a coaxial current probe placed at $\phi=0$, since $\sin\nu\varphi|_{\varphi=0}\equiv0$. Alternatively, a pair of differential current probes could be symmetrically placed with respect to the angular bisector (to emulate the unit doublet function of $\delta'(\phi)$) so that the CM-modes could be sufficiently excited. Otherwise, a centered aperture etched on the ground plane can be incorporated to excite the sine-dependent resonant modes. The case illustrated by Eq. (1.13) is exactly interchanged as that of Eq. (1.12). According to the properties of eigenmodes, the order of all usable modes can be felicitously excited and tuned upon using the gauged magnetic dipole. In this manner, the "mode gauged" or "mode synthesis" functionality (Lu et al. 2019) with a set of usable resonant mode tables has been realized, and it has been successfully employed to design compact dual-mode resonant wideband patch antennas (Lu et al. 2018), high-gain dual-mode resonant wideband patch antennas (Lu et al. 2017), dual-mode resonant tilted circularly polarized patch antenna (Yu et al. 2020), and triple-mode resonant null frequency scanning patch antenna (Wu et al. 2020). As will be seen, the mode gauged design approach to multi-mode resonant, circular sector patch antennas can be successfully developed based upon the generalized odd-even mode theory. More technical details will be discussed in Chapter 5.

1.2.3 Discussions and Remarks

According to Eqs. (1.3)–(1.5), and a series of typical examples tabulated in Table 1.1, several useful properties with generality and universal design guidelines for multi-mode resonant antenna designs can be concluded.

1. For an arbitrary resonator/antenna, the symmetry of eigenmodes is only dependent on the boundary condition of itself, and it is independent of the excitation. For the simplest, straight symmetric dipoles with specific boundary conditions, the even- and odd-symmetric eigenmodes should obey the "cosine-sine/sine-cosine alternating discipline (Leontovich and Levin 1944, Lu et al. 2019)".

2. The "symmetry" and "order" of eigenmodes are independent of each other. That's to say, the odd-order eigenmodes can exhibit even-symmetry (i.e., dipole), as well as odd-symmetry (i.e., loop), and so as to the even-order eigenmodes.

3. The excitation and combination of eigenmodes are dependent on the property of excitation source function. When the antenna is symmetrically excited, an even-symmetric source function can excite even-symmetric eigenmodes only, so as to the odd-symmetric source.

4. When the symmetry/boundary condition of the excitation source function is matched to those of the eigenmodes, the symmetry/ mode matching may lead to sufficient excitement and resonance of the corresponding eigenmodes. If the symmetry/boundary condition of the excitation source function is mismatched to those of the eigenmodes, the symmetry/mode mismatching between "feed" and "antenna" may lead to suppression or insufficient excitation of the corresponding eigenmodes.

5. For linear antennas, the mode mismatching in (4) may cause reactions between the feed line and dipole. In this manner, undesired sheath currents would be excited on the outer surface of the feed cable (King 1943). The parasitic sheath currents would lead to all kinds of unbalanced phenomena (King 1943, Hu 1987), such as unpredicted distortion of radiation patterns, high cross-polarization levels, and hot cable effect (Schantz 2015). Therefore, "balancing devices" (Hu 1987) or "balance to unbalance transformers" (baluns) should be incorporated to match the operational modes, and to suppress the unbalance effect.

6. Mathematically, the unbalanced currents should be dominated by the cross-coupled terms in the interior Green's function, as shown in Eqs. (1.3)–(1.5).

The six basic properties mentioned above have been proven to be effective for providing generally useful design guidelines to multi-mode resonant antennas, as well as reasonable explanations to the unbalanced effect in antenna systems.

1.3 THE HISTORY OF MULTI-MODE RESONANT ANTENNAS

In the previous sections, we have established a general theoretical framework for multi-mode resonant antennas by introducing the generalized odd-even mode theory, which has been proven to be correct and effective by a series of typical elementary antennas. Let's continue to rediscover the history and theory of multi-mode resonant antennas in this section.

Indeed, the ultimate idea of multi-mode resonant dipoles was originally pioneered in (Leontovich and Levin 1944) and employed to analytically calculate the surface current distributions, as well as input impedance for straight dipoles under arbitrary excitations. In that era, most antennas were designed to resonate at their fundamental mode only. For a long time, the originally advanced multi-mode resonant dipole concept was rarely mentioned and seldom employed in wideband or broadband antenna design problems.

With the development of high-frequency radio technologies, various complicated, wideband antennas and antenna systems have been becoming more and more desirable. Since the 1950s, the multi-mode resonant antenna theory has been employed to instruct antenna system layouts in aircraft: Qualitatively, resonant current modes on an airframe can be classified into "symmetric mode" and "antisymmetric mode", respectively, as illustrated in Figure 1.5. According to the inherent resonant current modes on the fuselage, various high-frequency and ultrahigh-frequency communication antennas can be reasonably installed, and naturally "isolated" to coexist, owning to the orthogonality of such modes (Granger and Bolljahn 1955). It is interesting to find that the complicated "symmetric mode" and "antisymmetric mode" exhibit similar even- or odd-symmetry (with reference to the fuselage axis) as the simplest one that has been mathematically depicted by trigonometric functions in (Leontovich

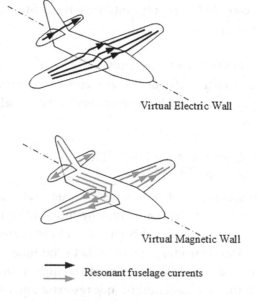

Virtual Electric Wall

Virtual Magnetic Wall

→ Resonant fuselage currents

FIGURE 1.5 Symmetric and antisymmetric resonant current modes of a complicated fuselage (Granger and Bolljahn 1955).

and Levin 1944). By comparing Figures 1.1, 1.2, and 1.5, it is seen that the "cosine-sine alternating discipline" (Lu et al. 2019) should be indeed the alternating symmetry of eigenmodes depicted in the generalized odd-even mode theory, and it should be versatile for arbitrary complex antennas or antenna systems with at least symmetry in one dimension.

Later on, the multi-mode resonant concept has also been specifically applied in some horn antennas designs (Goatley and Green 1956, Jennetti and Uda 1967, Clavin 1975) besides the complicated aircraft antennas: For regular waveguides, the resonant modes could be analytically solved so that they could be felicitously excited to generate circularly polarized radiation (Goatley and Green 1956), high gain (Jennetti and Uda 1967), and wideband equal E- and H-plane operations (Clavin 1975): As an example, a waveguide-fed slot antenna by the simultaneously exciting the H_{10} and H_{30} modes can lead to an increment of directivity of 75% over a wide frequency range of 2:1 (Jennetti and Uda 1967). However, limited by the computers' processor, memory, and effective numerical solvers at that era, the multi-mode resonant concept based upon "one radiator, multiple resonant modes" idea was still rarely applied for wideband antenna and complex antenna system designs. With the great progress of computational ability, and rapid development of high-frequency, wireless communication systems, high-performance, wideband antenna designs have recently been drawing more and more attention. To rapidly yield a prototype design with promising performance in the initial design step, generalized, physically insightful wideband antenna design approaches would be highly desired.

As is well known and widely recognized, using traveling-wave or frequency-independent antenna prototypes can effectively yield all kinds of wideband/broadband antenna designs (Schantz 2005). Since the 1950s, most broadband antennas have been designed and implemented by using the truncated traveling-wave or frequency-independent prototype antennas, i.e., bi-conical antennas (Goatley and Green 1956, Clavin 1975), bulbous antennas with smooth transitions and profiles (Schelkunoff and Friis 1952, Carrel 1958, Gibson 1979, Kraus 1988, Agrawall et al. 1998), periodical helix (Kraus 1948) and spiral antennas (Dyson 1959), and truncated self-complementary antennas (Rumsey 1966, Mushiake 1996). As is noted, only the principal resonant transverse electromagnetic mode is excited for the purpose of radiation in these cases.

The wideband/broadband characteristic of single-mode resonant, truncated traveling-wave antennas have been explained by the impedance frequency responses of infinity and truncated bi-conical antennas (Lu et al. 2019). For an infinity bi-conical antenna with a flared angle of $2\theta_0$, its input impedance should be angular-dependent only (Schelkunoff and Friis 1952), such that

$$R_i = 120 \ln \cot\left(\frac{\theta_0}{2}\right) \qquad (1.14)$$

The frequency-independent impedance characteristic depicted by Eq. (1.14) has been compared with the truncated case in (Lu et al. 2019). As is illustrated by Eq. (1.14), the input resistance of an ideally infinity bi-conical antenna can be indeed described by a Heaviside step function in the frequency domain (valid for all non-negative frequencies). It implies that such antenna should be non-causal that could never be physically realizable. This case is just analogous (but not equivalent) to one of the ideal, high-pass filters with a very similar brick-wall, amplitude frequency response (Schelkunoff and Friis 1952).

A causal and physically realizable case should be the truncated bi-conical antenna, in which both input resistance and reactance exhibit frequency-dependent, damped oscillations with ripples in magnitude around the Heaviside step function (indicating the input resistance R_i), and the horizontal real axis (indicating zero-reactance), respectively. This may lead to a multiple rippled frequency response in the reflection coefficient over an extremely wide frequency range. Intuitively, the residual, damping oscillated wideband characteristic should be inherited from the frequency-independent characteristic, as graphically illustrated in (Lu et al. 2019). Physically, it can also be explained by the superposition of propagated and reflected principal, transverse electromagnetic waves in truncated, finite frequency-independent antennas.

Mathematically, the wideband, damped oscillated impedance characteristic illustrated in (Lu et al. 2019) can be interpreted as the truncation of "continuous spectrum", "nonmodal solution" or "leaky mode" of the source-free wave equation (Marcuvitz 1956, Menzel 1978). Therefore, at root, the wideband antenna design approach based upon truncated traveling-wave or frequency-independent antenna should be mathematically distinctive to the multi-mode resonant one based upon the "discrete spectrum", "eigenmode solutions" or "resonant mode" (Marcuvitz 1956) of the source-free wave equation investigated in the previous sections. Further comparisons and discussions will be carried out in Chapter 7.

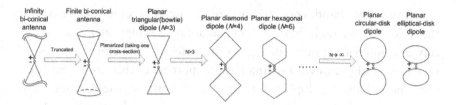

FIGURE 1.6   Evolution from frequency-independent to wideband dipole antennas. From infinity bi-conical antenna to wideband, circular/oval disk dipole antennas.

Figure 1.6 shows the evolution procedure from infinity frequency-independent bi-conical antenna to all kinds of wideband, disk-like dipoles. When an infinity bi-conical antenna is truncated, a wideband bi-conical antenna can be attained. Intersecting one cross section of the 3-D, bi-conical antenna may yield a planar, triangular (i.e., polygon with edge number of $N=3$) dipole, or a "bowtie" dipole (Schantz 2015). When the edge number $N>3$, it may yield more flat, disk dipoles with other polygonal shapes, e.g., diamond (square), pentagonal, hexagonal, etc., in further. When the edge number of N is approaching infinity, it would ultimately yield a circular disk dipole. Other smoothly shaped variants, i.e., elliptical, oval, water drop, etc., can be attained in the same way accordingly (Schelkunoff and Friis 1952, Honda et al. 1992, Liang et al. 2005). In this opinion, it can be recognized that the wideband, arbitrarily shaped flat disk dipole or alternatively, the Schelkunoff–Friis dipole (Simpson 2006), should originate and evolve from the traveling-wave, infinity bi-conical antenna, i.e., the simplest frequency-independent antenna.

In the fashion of engineering applications, as intensively illustrated in (Schantz 2015) and observed from Figure 1.6, it can be reasonably interpreted why "fatter", smooth silhouettes are always popular and widely used in broadband antenna designs. However, the elaborated "fatter" configuration may potentially incorporate higher structural complexity to design and manufacturing. For strongly resonant antennas such as slim dipoles, MPAs (Carver and Mink 1979, Pozar 1992), and dielectric resonator antennas (Long 1983, Luk and Leung 2003), the design approach based on truncated frequency-independent antennas will be no longer suitable. To maintain a relatively simple, slim configuration for resonant antennas, multiple resonant modes can be excited simultaneously and employed to form a broadened radiation bandwidth.

On the other hand, compared to conventional antennas resonating at the fundamental mode, high-order resonant modes may offer a new property and interesting performance to the antenna. Therefore, it is necessary to investigate novel antenna design approaches that are based upon the "one radiator, multiple resonant modes" concept: Multi-mode resonant antenna design approach would provide clear physical insights by precisely tuning and flexibly controlling the usable resonant modes. It would be beneficial in future wideband, high-performance antenna developments. In the following sections of this book, design approaches based on generalized odd-even mode theory and multi-mode resonant concept will be systematically introduced to typical resonant elementary antennas and compared to other design approaches. Relevant applications with miscellaneous examples will be raised and discussed in a comprehensive manner.

1.4 CONCLUDING REMARKS

In this chapter, we advance a generalized odd-even mode theory for multi-mode resonant antennas, and then rediscover the history and theory of multi-mode resonant antennas. Under the respective excitation of unit Dirac delta function and unit doublet function, the surface current/E-field distributions of multi-mode resonant antennas have been mathematically modeled by introducing even- and odd-symmetric interior Green's functions expanded in terms of even- and odd-symmetric eigenfunctions. Accordingly, concepts of DE-, CE-, DM-, CM-modes, and the degree of asymmetry/unbalance can be advanced to depict the excitation status of arbitrary multi-mode resonant antennas.

Compared to traditional antenna design theory (regardless of the mismatching and reactions between an antenna and its feed network), the generalized odd-even mode theory can possibly be employed to describe the relationship between antenna and feed network in a more general sense: Both mode-matching (balanced) and mode-mismatching (unbalanced) cases between antenna and source can be described. In this manner, usable resonant modes as well as the effective feeding schemes can be clearly identified, and intuitively determined so that useful design guidelines can be insightfully illustrated.

Furthermore, the unbalanced phenomena in antenna systems can be mathematically interpreted by the cross-coupled terms in the interior Green's function expanded in terms of eigenmodes, and physically explained owing to the concept of mode matching/mismatching between antennas and feed lines (King 1943, Geyi 2000). The proposed theory has been well validated by a series of typical antenna design examples. As will be proven, the generalized odd-even mode theory can provide a rigorous mathematical model and clear physical insights as well as general design guidelines for all kinds of multi-mode resonant elementary antennas. It lays down a unified theoretical framework for all resonant antenna design approaches. More rigorous mathematical interpretations for the generalized odd-even mode theory can be referenced to the odd/even symmetry of eigenfunctions of the Laplacian in two dimensions that has been discussed in (Kuttler and Sigillito 1984). In general, the generalized odd-even mode theory is more easily understandable than the rigorous method based on intergo-differential equation (Leontovich and Levin 1944). Therefore, it can be conveniently applied in both practical engineering and pedagogical activities.

In the following chapters, the generalized odd-even mode theory and the concept of "one radiator, multiple resonant modes" will be employed to design all kinds of multi-mode resonant elementary antennas at the simplest, basic form, as well as arrays, with diverse performance and different functionalities for all kinds of applications.

CHAPTER 2

Multi-Mode Resonant Electric Dipole Antennas

2.1 BRIEF HISTORY OF RESONANT ELECTRIC DIPOLE ANTENNAS

The evolution of traveling-wave, bi-conical dipoles for wideband or impulse radios has been extensively introduced in (Schantz 2015) and briefly summarized in Chapter 1, thus we'd like to introduce the brief history in the opinion of "resonant electric dipoles" herein. The milestones of resonant electric dipole antennas have been graphically illustrated in Figure 2.1. Since Hertz pioneered the spark-gap experiment and validated the existence of electromagnetic waves in 1,888 (Schantz 2005), electric dipoles have been the most classical but the most popular, and widely used antennas, as shown in Figure 2.1a. Hertz also pioneered the concept that an adjacent parabolic cylindrical metallic reflector can make radio frequency energy more focused to generate directive radiations (Schantz 2015), which evolved into the conceptual design of "corner reflector antenna" later on (Kraus 1940), as illustrated in Figure 2.1b. In the early era of antenna technology, linear electric dipole antennas with high directivity have been extensively studied.

Figure 2.1c shows the milestone of Yagi–Uda antenna (Yagi 1928). In that pioneer work (Yagi 1928), Yagi advanced the concept of "wave canal" by properly incorporating a reflector and multiple directors to

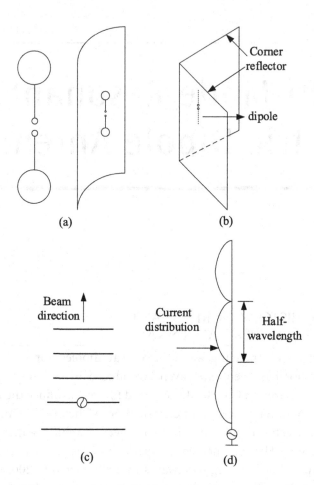

FIGURE 2.1 Milestones of resonant electric dipole antennas: (a) the Hertzian dipole invented in the spark-gap era (Schantz 2015), (b) the corner reflector antenna Kraus (1940), (c) the Yagi–Uda antennas (Yagi 1928), and (d) the Franklin dipole array antenna (Brown 1936).

a resonant dipole, and then yielding linear dipole antennas with high directivity (Yagi 1928) , for the first time. Such directive dipole antennas with simple, linear reflectors have been designed, manufactured in mass, and widely used in all kinds of wireless applications from high-frequency to ultra-high-frequency bands since the 1940s. One of the most popular applications of Yagi–Uda antenna should be the household receiving antenna for frequency modulation and television broadcasting.

Figure 2.1d shows the Franklin antenna (Brown 1936). It is composed of multiple vertical dipoles fed in series for narrowing down elevation radiation patterns and improving field intensity. Since then, all kinds of dipole/monopole antennas/arrays with arbitrary loads and parasitic elements (Brown 1937) have been developed and studied in parallel. Meanwhile, the analytical approaches to linear electric dipole antennas have been established and developed to study the current distribution and input impedance under arbitrary excitations (Leontovich and Levin 1944, Tai 1948, King 1956). With the development of modern computers, mature of numerical computational approaches, and computer-aided design tools, countless electric dipole/monopole antennas have been developed and implemented since then. To date, the simplest, linear electric dipole antenna and its variants are still widely applied in all kinds of wireless systems. In most of the previous works, the electric dipole antennas under investigation are supposed to be single-mode resonant. In this chapter, we will study the electric dipole antennas in the opinion of multiple-mode resonance.

2.2 EIGENMODES OF THE SIMPLEST, RESONANT STRAIGHT ELECTRIC DIPOLE

Let's begin with the simplest straight, time-harmonic, loss-less electric dipole in vacuum (Leontovich and Levin 1944) to illustrate the multi-mode resonant electric dipole antenna theory. Suppose each arm of the dipole is $L/2$ in length, with its axis coinciding with the z-axis, and both ends are left opened. In addition, the dipole should be slim enough, such that its radius $a \ll L$. To determine the eigenmodes of the dipole, only the source-free, homogeneous equation is considered. Thus, the eigenmode current distributions of $I_d(z)$ should obey the telegrapher's equation under specific boundary conditions, as indicated by Eq. (2.1). (Leontovich and Levin 1944, Lu et al. 2018, Lu et al. 2019).

$$\left(\frac{d^2}{dz^2}+k_z^2\right)I_d(z)=0, I_d\left(-\frac{L}{2}\right)=I_d\left(\frac{L}{2}\right)=0 \qquad (2.1)$$

The general solution for Eq. (2.1) can be represented by the linear combinations of sine and cosine functions in Eq. (2.2), where A, B are undetermined coefficients. Substituting the boundary condition in Eq. (2.1) into the general solution of Eq. (2.2), and for non-trivial solutions of A and B, the determinant of the matrix equation should vanish, such that

$$I_d(z) = A\sin k_z z + B\cos k_z z \Rightarrow \begin{vmatrix} \sin k_z \dfrac{L}{2} & \cos k_z \dfrac{L}{2} \\[2ex] -\sin k_z \dfrac{L}{2} & \cos k_z \dfrac{L}{2} \end{vmatrix}$$

$$= 0 \Rightarrow \sin k_z \frac{L}{2}\cos k_z \frac{L}{2} = 0 \tag{2.2}$$

Thus, the eigenvalue (i.e., wave number) k_z, full set of eigenmode current functions, and characteristic wavelength λ_z can be solved as

$$\sin k_z \frac{L}{2} = 0 \Rightarrow k_z = \left\{ \frac{2n\pi}{L} \right\}, n = 1, 2, \ldots$$

$$\cos k_z \frac{L}{2} = 0 \Rightarrow k_z = \left\{ \frac{(2n+1)\pi}{L} \right\}, n = 0, 1, 2, \ldots \tag{2.3a}$$

$$I_d(z) = \left\{ \cos\frac{\pi}{L}z, \sin\frac{2\pi}{L}z, \cos\frac{3\pi}{L}z, \sin\frac{4\pi}{L}z, \ldots \sin\frac{2n\pi}{L}z, \right.$$

$$\left. \cos\frac{(2n+1)\pi}{L}z, \ldots n = 0, 1, 2, \ldots \right\} \tag{2.3b}$$

$$\lambda_z = \frac{2L}{n}, n = 1, 2, \ldots$$

$$P(\theta) = \begin{cases} \dfrac{\cos\left(\dfrac{n\pi}{2}\cos\theta\right)}{\sin\theta}, & n = 1, 3, 5, 7, odd \\[4ex] \dfrac{\sin\left(\dfrac{n\pi}{2}\cos\theta\right)}{\sin\theta}, & n = 2, 4, 6, 8, even \end{cases} \tag{2.4}$$

As can be seen from Eq. (2.3a) and (2.3b), the length L of a resonant dipole must be the integer multiples of the one-half characteristic wavelength λ_z (Leontovich and Levin 1944). Using the eigenmode current functions, the general expressions of the characteristic radiation patterns (Lu et al. 2018) can be readily calculated by Eq. (2.4). The characteristic radiation patterns of the previous four resonant modes have been analyzed and presented in Figure 2.2. As can be seen, the odd-order (i.e., one half-, 1.5-wavelength, etc.) dipole modes exhibit identical E-plane patterns to those predicted by classical dipole theory, while the even-order (i.e., full-, double-wavelength, etc.) ones' E-plane patterns exhibit nulls in the broadside directions of $\theta=\pm90°$. These results match well with the ones presented in a classical reference Carter et al. (1931). As discussed in Chapter 1, all even-order, CE-modes within the dipole can rarely be excited under centered, even-symmetrical excitation. Alternatively, they could be partially or sufficiently excited by using off-center or dual-port feeds.

Therefore, multi-mode resonant dipole antennas can be accordingly designed by properly exciting the current's eigenmodes shown in Eq. (2.3b). In engineering practice, there are many ways to manipulate usable eigenmodes. In the following sections, different multi-mode resonant dipole antenna designs will be presented and discussed.

2.3 DESIGN OF 1-D MULTI-MODE RESONANT, CENTER-FED STRAIGHT DIPOLE ANTENNA

For brevity, the simplest, z-oriented symmetric electric dipole with loaded stubs in Figure 2.3 is studied as the first example. The dipole is fed at the center so as to excite the even-symmetric, odd-order Differential Electric (DE) modes only. As can be seen from Figure 2.3, a pair of x-oriented tuning stubs is symmetrically added near the two respective electric current nodes of the 1.5-wavelength (i.e., the third order) resonant mode. As the stubs' length increases, the resonant frequency of the 1.5-wavelength mode could be gradually tuned down while that of the half-wavelength mode would be barely affected (Zhu et al. 2005).

Wideband dual-mode resonant dipole antennas have been accordingly developed (Lu, Zhu et al. 2017, Kuo et al. 2010, Hsu and Huang 2012) based upon the dual-mode resonant idea illustrated in Figure 2.3. Both unilaterally and bilaterally stub-loaded cases shown in Figure 2.4 have been discussed Lu, Zhu et al. (2017). When there is only a single pair of x-oriented stub unilaterally introduced, the antenna is asymmetric about the z-axis. As discussed in Chapter 1, a symmetric, z-oriented dipole under pure DE-mode excitation should exhibit natural infinite electric wall boundary conditions in xy-plane (i.e., the H-plane), and infinite magnetic wall boundary

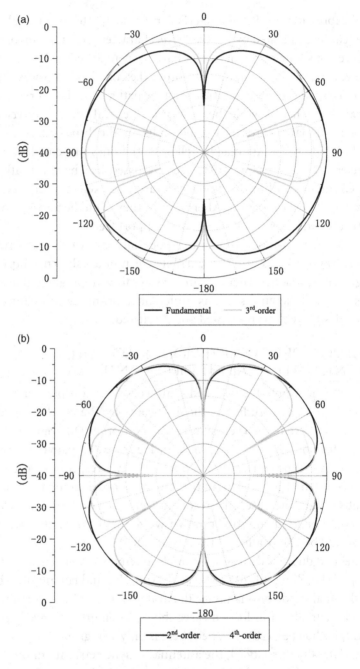

FIGURE 2.2   Normalized characteristic radiation patterns of the first four resonant modes within a symmetric dipole: (a) half-wavelength and 1.5-wavelength mode and (b) full-wavelength and double-wavelength mode.

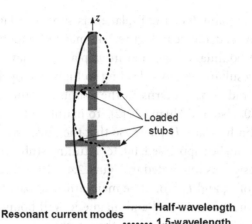

Resonant current modes
—— Half-wavelength
······· 1.5-wavelength

FIGURE 2.3 Conceptual illustration to dual-mode resonant, vertical dipole antenna bilaterally loaded by tuning stubs (Lu, Zhu et al. 2017, Lu et al. 2019).

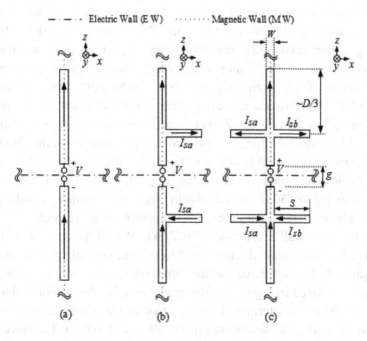

FIGURE 2.4 Surface current distributions of the center-fed, stub-loaded dipole (Lu, Zhu et al. 2017): (a) the natural boundary condition of an ideal dipole for comparison, (b) the unilateral loaded case, and (c) the bilateral loaded case.

conditions in yz-plane (i.e., the E-plane), as shown in Figure 2.4a. Thus, the unilaterally loaded case in Figure 2.4b may not maintain the inherent magnetic wall boundary condition in yz-plane. It may distort the symmetry of the resultant antenna, leading to high cross-polarization levels and distorted radiation patterns in the azimuth plane, as illustrated in (Kuo et al 2010, Hsu and Huang 2012). To maintain the dipole's inherent symmetry in both E- and H-planes so that the undesired cross-polarized components could be suppressed, bilateral tuning stubs would be required in practical design, as illustrated in Figure 2.4c. Therefore, the horizontal stub currents of I_{sa} and I_{sb} would be maintained to balance and obey the respective natural electric wall and magnetic wall boundary conditions, so that the cross-polarization level in the azimuth plane can be effectively reduced.

A printed, bilateral stub-loaded dipole antenna is designed and studied. Unlike the full-metal case in (Lu, Zhu et al. 2017), the printed bilateral case is designed on a dielectric substrate with relative permittivity $\varepsilon_r = 2.65$ and thickness $h = 1$ mm at 3 GHz band. The antenna is designed on one side of the substrate to yield a single-layered printed dipole with initial parameters of $S = 12.2$ mm, $g = 0.4$ mm, $D = 56.0$ mm, and $W = 9.0$ mm. Three parametric studied curves on W are plotted in Figure 2.5a and b. As can be seen, when $W = 9.0$ mm, a relatively flattened dual-mode resonant reflection coefficient frequency response can be satisfactorily attained. Figure 2.5b illustrates the input impedance on Smith Chart. Compared to the slim, full-metal case (Lu, Zhu et al. 2017) with average input resistance up to a few hundred Ohms, the printed dipole's input resistance is much smaller and closer to 50 Ω over a wide bandwidth, owning to the loaded effect of dielectric substrate. Therefore, the printed dipole tends to be more easily matched to a 50 Ω coaxial cable in direct, without using a wider profile Lu, Zhu et al. (2017) or an external impedance transformer.

Similar to the full-metal case in Lu, Zhu et al. (2017), the printed dipole is also adhesively bonded and excited by a 50 Ω coaxial cable, as shown in Figure 2.6. The simulated and measured reflection coefficients are plotted and compared in Figure 2.7. The impedance bandwidth for $|S_{11}|$ lower than -10 dB can be determined from 2.2 to 4.2 GHz, which is about 62.5%, and wider to the 49.7% reported in Lu, Zhu et al. (2017). Therefore, the loaded dielectric substrate should be beneficial for enhancing the impedance bandwidth of a dual-mode resonant dipole at the simplest, basic configuration.

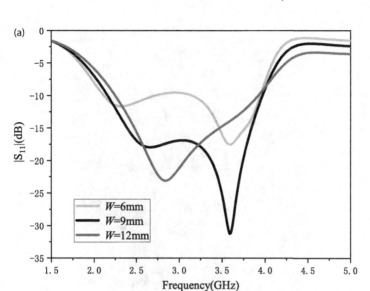

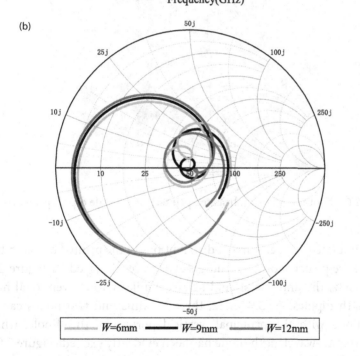

FIGURE 2.5 Reflection coefficient of the printed stub-loaded dipoles with respect to width W ($S=12.2\,\text{mm}$, $D=56\,\text{mm}$, $g=0.4\,\text{mm}$): (a) reflection coefficients frequency response with reference to different intrinsic impedance (in dB) and (b) input impedance curves on Smith chart.

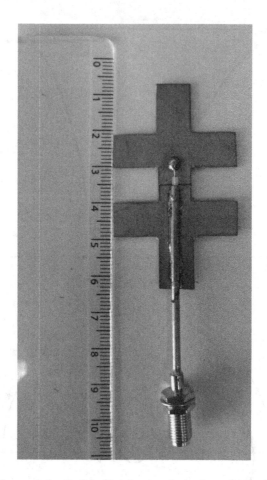

FIGURE 2.6 Photograph of a fabricated printed, dual-mode dipole prototype.

The simulated surface current distributions of the printed dipole at two resonant frequencies of 2.59 and 3.92 GHz are displayed in Figure 2.8. At 2.59 GHz, the antenna should behave similarly to conventional half-wavelength dipoles. At 3.92 GHz, three maxima and two nodes can be found along with the principal, vertical portion of the dipole, which implies the 1.5-wavelength mode has been evidently excited (Figure 2.9).

The E-(yz-plane) and H-plane (xy-plane) radiation patterns at two resonant frequencies of 2.59 and 3.92 GHz are simulated, measured, and compared with each other. As can be seen, the antenna typically exhibits an "8" shape, E-plane pattern and nearly "O" shape, H-plane one. The non-uniformity of H-plane pattern is about 3.2 dB, which is slightly higher

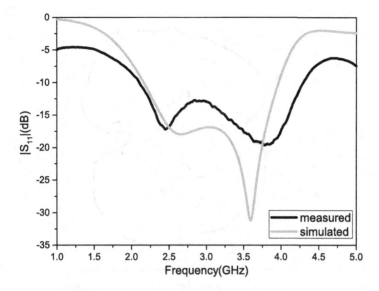

FIGURE 2.7   Simulated and measured reflection coefficients of the printed dual-mode resonant dipole.

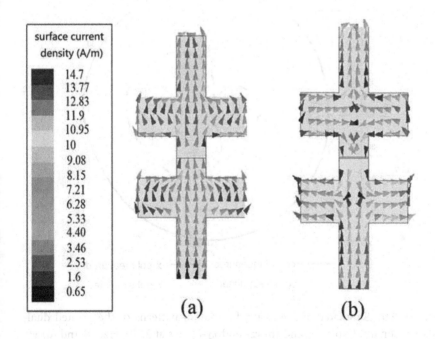

FIGURE 2.8   Simulated surface current distributions of the printed dual-mode resonant dipole: (a) 2.59 GHz and (b) 3.92 GHz.

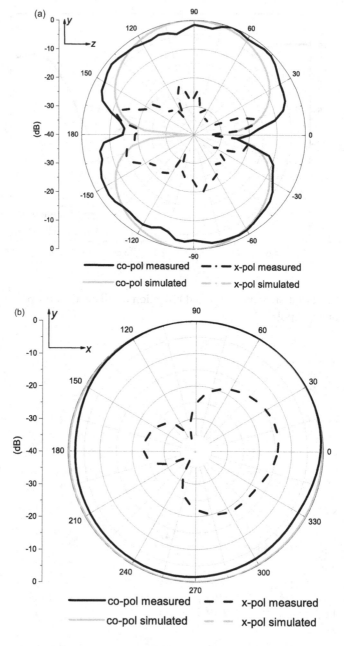

FIGURE 2.9 Simulated and measured radiation patterns of the printed dual-mode resonant dipole: (a) and (b) yz- and xy-planes at 2.59 GHz; (c) and (d) yz- and xy-planes at 3.92 GHz.

(*Continued*)

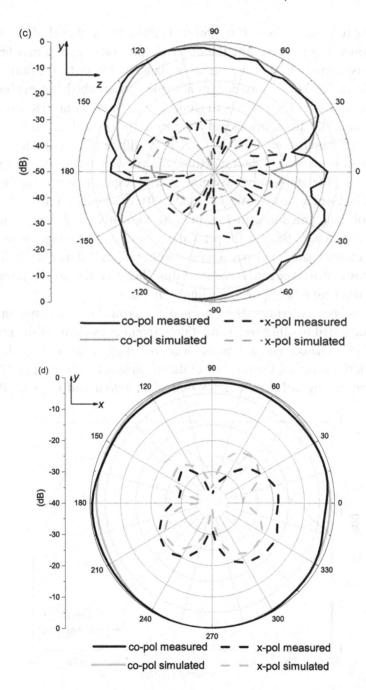

FIGURE 2.9 (*Continued*) Simulated and measured radiation patterns of the printed dual-mode resonant dipole: (a) and (b) *yz*- and *xy*-planes at 2.59 GHz; (c) and (d) *yz*- and *xy*-planes at 3.92 GHz.

than the full-metal case in (Lu, Zhu et al. 2017). The higher H-plane non-uniformity may be caused by the dielectric substrate, since it may bring minor asymmetry to the antenna in xy-plane. At 3.92 GHz, the radiation pattern exhibits a triple-lobe doughnut shape instead, which is contributed by the radiation of 1.5-wavelength mode. Unlike the ideal case (Kraus and Marhefka 2002), the spurious radiation from the vertical feed cable cannot be fully suppressed, thus the directivity will be enhanced due to the superposition of two resonant radiation-contributed modes, and the parasitic effects. Therefore, minor asymmetry with respect to y-axis and beam distortion (Schantz 2003, Guéguen et al. 2005) can be observed.

Finally, the antenna gain is simulated, measured, and compared in Figure 2.10. As can be seen, the printed dipole antenna exhibits a multiple mode resonant gain frequency response within the impedance bandwidth. The measured average gain is about 2.5 dBi. Therefore, the gain bandwidth can be determined as equal to the impedance one.

As has been experimentally validated, the printed dipole is systematically compared to other wideband dipole antennas in terms of the numbers of the radiator, resonant mode for contributing radiation, impedance bandwidth, gain, and the number of design parameters (Kuo et al. 2010, Hsu and Huang 2012, Kuo and Wong 2003, Tefiku and Grimes 2000,

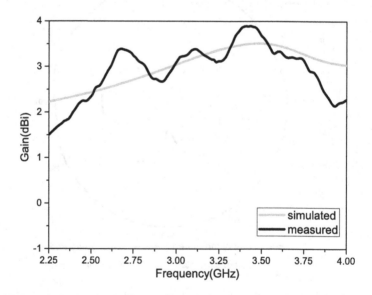

FIGURE 2.10   Simulated and measured gains of the printed dual-mode resonant dipole.

TABLE 2.1   Comparisons Between Wideband Dipoles (N.A. = Not Available)

References	Radiators/Resonant Mode for Radiation	Impedance Bandwidth (%)	Average Gain	Number of Design Parameters
Kuo et al. 2010	Dual/multiple	131.0	N.A.	5
Hsu and Huang 2012	Dual/dual	63.8	2.0 dBi	7
Kuo and Wong 2003	Dual/dual	23.5	1.5 dBi	10
Tefiku and Grimes 2000	Dual/dual	30.0	5.0 dBi	12
Behera and Harish 2012	Dual/dual	66.0	2.5 dBi	10
Le and Karasawa 2012	Single/single	95.5	3.3 dBi	10
Yang et al. 2012	Single/single	114.7	4.0 dBi	17
Lu et al. 2014	Dual/single	129.0	4.5 dBi	15
Wang et al. 2014	Single/single	113.0	8.2 dBi	32
Luo et al. 2019	Single/dual	11.2	4.0 dBi	18
Li and Li 2020	Single/dual	51.4	3.0 dBi	14
This work	**Single/dual**	**62.5**	**2.5 dBi**	**4**

Behera and Harish 2012, Yang et al. 2012, Wang et al. 2014, Le and Karasawa 2012, Lu et al. 2014, Luo et al. 2019, Li and Li 2020). As tabulated in Table 2.1, the printed dual-mode resonant dipole simultaneously uses half- and 1.5-wavelength modes for radiation, which is distinctive to its counterparts. The dipole antenna exhibits the simplest configuration with only four key design parameters. Therefore, the 1-D linear, straight dual-mode resonant dipole (Lu, Zhu et al. 2017) has been experimentally investigated and convinced as the simplest 1-D multi-mode resonant elementary antennas.

2.4 DESIGN OF 2-D MULTI-MODE RESONANT SECTORIAL ELECTRIC DIPOLES

In the previous section, a 1-D dual-mode resonant dipole antenna has been investigated. Actually, the 1-D multi-mode resonant concept can be extended to more generalized, 2-D multi-mode resonant dipoles. In this section, the multi-mode resonant behavior of 2-D, planar current sheets will be discussed.

Upon using the mode gauged design approach (Lu et al. 2018, Lu et al. 2019) to magnetic dipoles, a 1-D multi-mode resonant gauged electric dipole can be introduced and evolved into a 2-D multi-mode resonant circular sectorial dipole (Zhao et al. 2021), and Transversal Electric (TE) mode would be excited instead of Transversal Magnetic (TM) ones. The case in Figure 2.11a can be treated as the complementary case discussed in

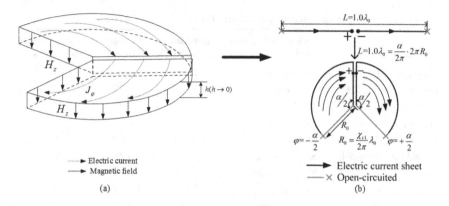

FIGURE 2.11 Evolution from a 1-D straight gauged electric dipole into a TE-mode resonant, 2-D circular sector dipole (Zhao et al. 2021).

(Lu et al. 2018). As can be seen from Figure 2.11b, the periphery of circular sector dipole should be set as the integer multiples of one half-wavelength, with both radii open-circuited.

As an example, we choose the length of the prototype dipole as $L = 1.0\lambda$, where λ denotes the wavelength of the center frequency. Using the mode gauged curve presented in Zhao et al. (2021), usable resonant TE modes, flared angle α and radius R_0 can be determined. Then, a full-wavelength, 2-D dual-mode resonant circular sector dipole antenna can be designed. At the first glance, the sectorial dipole in Figure 2.11 seems to be similar to a conventional bow-tie dipole. In the following section, we will set out to clarify the difference between them.

First of all, the bow-tie dipole originates from the frequency-independent, i-conical antenna operating at the transversal electromagnetic mode Schelkunoff and Friis (1952): A cross-section of bi-conical antenna may yield a bow-tie dipole Schantz (2015). When the conical/bow-tie dipole is truncated, damped-oscillated, wideband characteristic can be attained, and such wideband characteristic can be recognized as the residual of the frequency-independent characteristic (Lu et al. 2019) that is dominated by the reflection behavior of the principal transversal electromagnetic wave (Schelkunoff and Friis 1952, Schantz 2015, Kraus 1988) propagating in ideally, infinity frequency-independent antennas. The wideband bow-tie dipole antenna should be a truncated frequency-independent, traveling-wave (or leaky-wave) antenna, rather than a resonant one. Once truncated, the current distribution (Schelkunoff and Friis

1952, Kraus 1988) of a yielded bow-tie dipole antenna with an apex angle of α (with its axis coinciding with the z-axis) can be presented by

$$I(r,\theta) = I_0 \cos(kr)\sin\theta \tag{2.5}$$

As is seen from Eq. (2.5), the currents on a bow-tie dipole would tend to concentrate at the edges ($\theta = \pm\ \alpha/2$), with weakened currents to be distributed near the symmetric axis (for small θ), which was originally discovered by Hertz and re-described in Schantz (2015). Therefore, the current distribution on the edge ($\theta = \pm\ \alpha/2$) could be approximately simplified as

$$I(r) = I_0 \sin(\alpha/2)\cos(kr) \tag{2.6}$$

The expression in Eq. (2.6) well explains why the truncated bow-tie dipole could be approximately treated as a 1-D resonant dipole in most engineering designs. This is also the reason that the bi-conical antennas can be meshed and built as grid- or cage-dipoles in many low-frequency cases (Schantz 2015, Kraus 1988). In addition, Eq. (2.6) also well indicates that the bow-tie antenna should be fed at its apex (i.e., $r=0$) so that all cosine-dependent resonant modes could be sufficiently excited, just as the 1-D straight dipole discussed in Section 2.2.

While in our design approach, the situation is quite different. For facilitating theoretical analysis, we can start our discussion from the horizontal, 2-D sectorial dipole placed in xy-plane, as shown in Figures 2.11 and 2.12a. The dipole should be symmetrically fed at its angular bisector, i.e., the x-axis. Suppose the sectorial sheet is thin with finite thickness, with circumference short-circuited, and radii open-circuited. Thus the TE resonant modes can be analogously deduced from the longitudinal magnetic field of H_z, which is complementary to the case of TM mode resonant microstrip patch antennas (Lu et al. 2018). Eq. (2.7) shows the general expression of the electric current density distribution of the 2-D sectorial dipole (Lu et al. 2019). As is found, the 2-D resonant sectorial current sheet exhibits current distribution that is dominated by the Fourier-Bessel series, which is distinctive to that one shown in Eq. (2.5) to Eq. (2.6). Eq. (2.7b) and (2.7c) also clearly indicate that the feed position should not locate at the apex of the antenna (i.e., $\rho=0$), otherwise, all resonant TE modes would never be excited for $J_{n\pi/\alpha}(k\rho) \equiv 0$ ($n \neq 0$, $\rho=0$).

Thus the resonance and radiation behavior of the sectorial dipole should be distinctive from the ones of the bow-tie dipole.

Then, a 2-D sectorial dipole with flared angle $\alpha = 240°$ is designed to study its radiation characteristics under fundamental mode (i.e., $TE_{3/4,1}$ mode) resonance. The sectorial dipole is designed to operate at the center frequency of 2.4 GHz. The radius of the dipole can be approximately determined as $R_0 = 29.8$ mm using the formulas presented in Lu et al. (2018). Using the numerical integration of Eq. (2.8a) and Eq. (2.8b), the principally polarized radiation patterns in both E-plane (i.e., xy-plane) and H-plane (i.e., xz-plane) can be predicted, respectively, where k is the wave number, $k_{3/4,1}$ is the characteristic wave number for the resonant $TE_{3/4,1}$ mode, R_0 is the radius of the sectorial dipole, $I_{3/4,1}$ is the magnitude of the electric current under the fundamental mode resonance, and $J_{3/4}(\cdot)$ is the 3/4-order Bessel function of the first kind.

$$\left(\nabla^2 + k_{nm}^2 \right) H_z = 0$$

$$\vec{J}_\phi \left(\rho, \phi \right) = \hat{\rho} \times \vec{H}_z \left(\rho, \phi \right)$$

$$\frac{\partial H_z}{\partial \rho} \big|_{\rho=a} = 0, \vec{J}_\phi \left(\rho, \frac{\alpha}{2} \right) = \vec{J}_\phi \left(\rho, -\frac{\alpha}{2} \right) = 0 \tag{2.7a}$$

$$H_z = A_{nm} J_{\frac{n\pi}{\alpha}} \left(k\rho \right) \, \overset{sin}{sin} \frac{n\pi}{\alpha} \phi \tag{2.7b}$$

$$J_\phi \left(\rho, \phi \right) = \sum_{n=1}^{odd} \sum_{m=1}^{\infty} \frac{I_{\frac{n\pi}{\alpha},m} J_{\frac{n\pi}{\alpha}} \left(k\rho \right) \cos\left(\frac{n\pi}{\alpha} \phi \right)}{k^2 - k_{\frac{n\pi}{\alpha},m}^2} \tag{2.7c}$$

$$E_{\phi-xy} = \frac{I_{\frac{3}{4}-1} e^{-jkr}}{r} \int_0^{R_0} \rho' J_{\frac{3}{4}} (k\rho') d\rho' \int_{\frac{2\pi}{3}}^{\frac{2\pi}{3}} \cos\phi \cos(\frac{3}{4}\phi') e^{jk\rho'\cos(\phi-\phi')} d\phi' \tag{2.8a}$$

$$E_{\phi-zx} = \frac{I_{\frac{3}{4}-1} e^{-jkr}}{r} \int_0^{R_0} \rho' J_{\frac{3}{4}} (k\rho') d\rho' \int_{\frac{2\pi}{3}}^{\frac{2\pi}{3}} \cos(\frac{3}{4}\phi') e^{jk\rho'\sin\theta\cos(\phi-\phi')} d\phi' \tag{2.8b}$$

The normalized radiation patterns under $TE_{3/4,1}$ mode resonance are numerically simulated by employing Ansys High Frequency Structure Simulator (HFSS)™. In the simulation, the excited port is exactly set at $\rho = R_0$. As can be seen from Figure 2.13, the numerical and theoretical radiation patterns agree reasonably well with each other. The $TE_{3/4,1}$ mode resonant sectorial dipole exhibits an omnidirectional radiation pattern with widened E-plane pattern, just as conventional linear dipole and its 270° counterpart in (Zhao et al. 2021) behave.

In practice, the sectorial dipole should be placed within the elevation plane, so that it can be conveniently fed and supported by a vertical coaxial cable, as shown in Figure 2.12b. Peripherally, we can incorporate a pair of open-circuited, tuning stubs with length L_S and width W_S at $\beta = \pm 40°$, for resonant $TE_{9/4,1}$ mode excitation (Lu et al. 2017, Zhao et al. 2021), and to achieve a dual-mode resonant wideband characteristic. Empirically, we should approximately set L_S and W_S equal to one-quarter and one-tenth wavelength of $TE_{9/4,1}$ mode, respectively (Lu and Zhu 2015b, Guo et al. 2017). Symmetrically, a coaxial cable with a dummy portion is adhesively bonded to feed the antenna, as shown in Figure 2.14 (Wolosinski et al. 2019, Vorobyov et al. 2005). The initial design parameters are $R_0 = 29.8\,mm$, $L_S = 16.4\,mm$, $W_S = 6.4\,mm$, and $\beta = 40°$.

Parametric studies are carried out to attain optimal designs. As shown in Figure 2.15a–c, when a pair of L-shape stubs is loaded near the current distributions' node of $TE_{9/4,1}$ mode, the $TE_{9/4,1}$ mode can be excited and tuned to merge with the fundamental mode, to form a dual-mode resonant frequency response, just like the 1-D case in (Lu, Zhu et al. 2017) behaves. Accordingly, optimal parameters can be attained and listed in

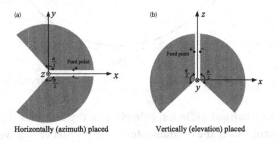

FIGURE 2.12 The evolved 2-D sectorial dipole with flared angle α placed in different planes: (a) azimuth plane and (b) elevation plane.

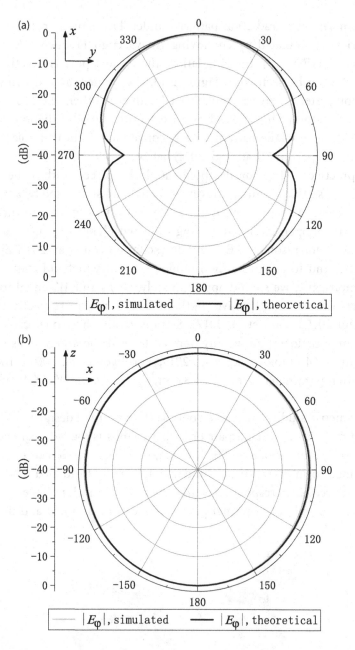

FIGURE 2.13 Normalized radiation behaviors of the sectorial dipole placed in an azimuthal plane under the fundamental mode resonance: (a) xy-plane and (b) xz-plane.

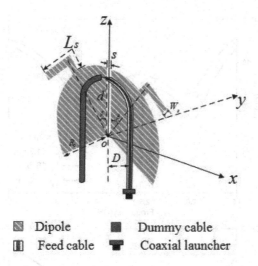

Dipole   Dummy cable
Feed cable   Coaxial launcher

FIGURE 2.14 Configuration of the wideband sectorial dipole antenna with coaxial cable feed.

Table 2.2. As can be seen, the initial and optimal results agree quite well, and it validates the design approach (Lu, Zhu et al. 2017, Zhao et al. 2021). Figure 2.16 shows the photograph of a prototype made of 0.2-mm thick, pure copper sheets and fed by a rigid, 50 Ω coaxial cable.

As can be seen from Figure 2.17, the measured reflection coefficient coincides quite well with the simulated one. The −10 dB impedance bandwidth is from 1.65 to 3.60 GHz, i.e., 74.3% in fraction. Figure 2.18 plots the simulated surface electric current distributions at 2.0 and 3.1 GHz. Intuitively, $TE_{3/4,1}$ mode at 2.2 GHz and $TE_{9/4,1}$ mode at 3.1 GHz can be excited in sufficient.

The radiation patterns at 2.2 and 3.1 GHz are measured and compared to the simulated ones. Figure 2.19 shows that all measured patterns match well with the simulated ones. The half-power beamwidth (HPBW) of the E-plane at 2.0 and 3.1 GHz is 112° and 70°, respectively. The current distributions on the curved periphery may cause a widened E-plane radiation pattern, as discussed in (Sun et al. 2018, Luo et al. 2017, Hamid et al. 1970), and this is also valid for a 2-D sectorial dipole. For low-frequency patterns, the visible discrepancy can be seen to occur in the directions from $\theta = 120°$ to 170°. Such discrepancy can be mitigated

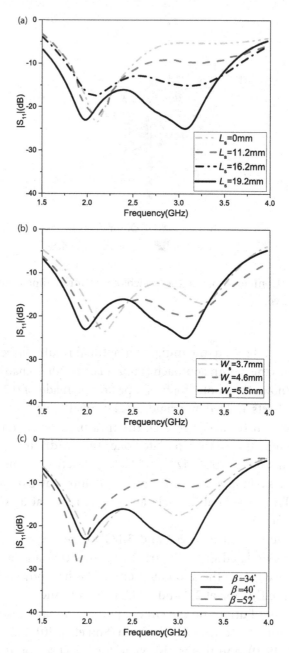

FIGURE 2.15 Parametric studies on the tuning stub: (a) L_S ($W_{S=}5.5$ mm, and $\beta = 40°$), (b) W_S ($L_{S=}19.2$mm, and $\beta = 40°$), and (c) β ($L_{S=}19.2$mm, and $W_S = 5.5$ mm).

TABLE 2.2   Parameters of the Proposed Antennas ($\alpha = 240°$)

	Parameters	
	Initial	simulated
R_0 (mm)	29.8	29.1
L_S (mm)	16.2	19.2
W_S (mm)	6.4	5.5
β (deg.)	40°	43°
s (mm)	1.4	
d (mm)	25.0	
D (mm)	10.0	

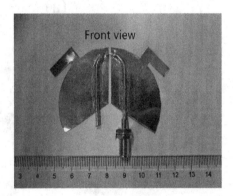

FIGURE 2.16   The photograph of a fabricated 2-D, dual-mode resonant sectorial dipole prototype.

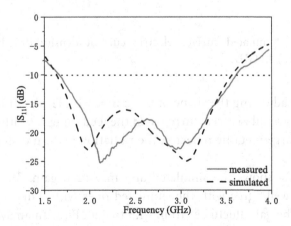

FIGURE 2.17   Simulated and measured reflection coefficients of the 2-D, dual-mode resonant sectorial dipole prototype.

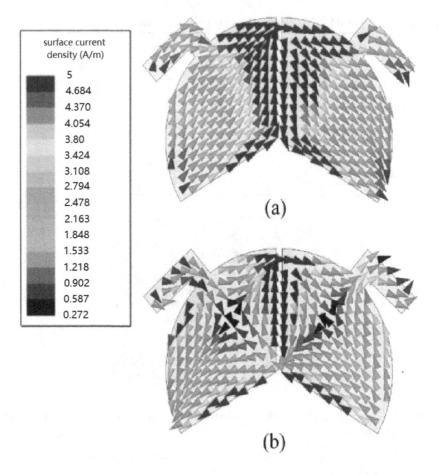

FIGURE 2.18 Simulated surface electric current density distributions: (a) 2.0 GHz and (b) 3.1 GHz.

by carefully adjusting the shape of the dummy cable, so as to fully suppress the stray, unbalanced currents on the feed cable's sheath. By incorporating a large metallic reflector, the radiation pattern distortions can be mitigated.

Figure 2.20 shows the simulated and measured gain. The measured gains fairly well agree with the simulated ones with a discrepancy less than 1 dB. The gain fluctuates from −1.3 to 4.9 dBi, with an average gain of about 2.0 dBi. As explained in Zhao et al. (2021), such characteristics are primarily caused by the mode mismatching between antenna and feed line (Schantz 2015, King 1943). Fortunately, a backed metallic reflector

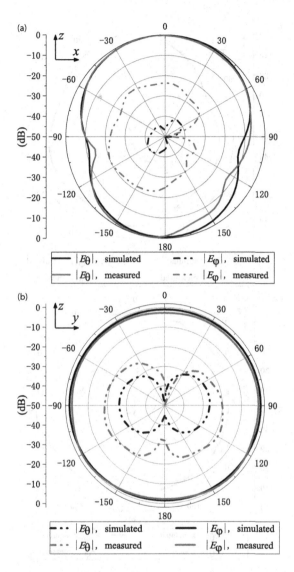

FIGURE 2.19 Simulated and measured normalized radiation patterns: (a) and (b) xz- and zy-planes at 2.0 GHz;

(*Continued*)

(Wolosinski et al. 2019, Zhao et al. 2021) would be helpful to realize a wideband, less-fluctuated gain performance.

Then, a prototype antenna mounted above a planar metallic reflector is studied, as shown in Figure 2.21a. The antenna height and diameter of

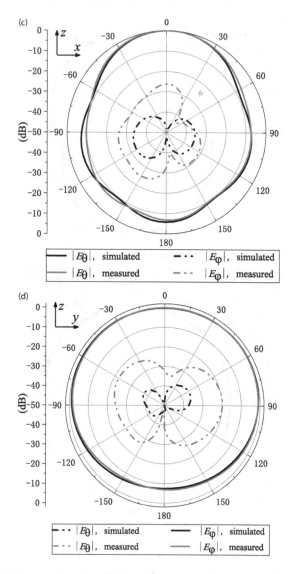

FIGURE 2.19 (*Continued*) (c) and (d) *xz*- and *zy*-planes at 3.1 GHz.

the circular reflector are identical to those presented in Zhao et al. (2021), and final parameters are tabulated in Table 2.3. Figure 2.21b shows the photograph of a fabricated prototype. Figure 2.22 illustrates the simulated and measured reflection coefficients, and the impedance bandwidth for $|S_{11}|$ lower than −10 dB can be determined 1.69 to 3.01 GHz, i.e., 56.2%

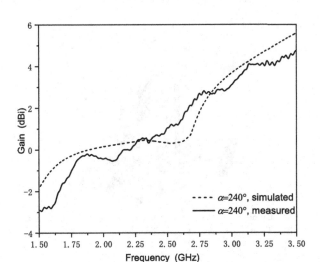

FIGURE 2.20   Simulated and measured radiation gains for the 2-D, dual-mode resonant sectorial dipole.

in fraction and slightly narrower than the one without reflector case (Figure 2.23).

By incorporating a metallic reflector, HPBW of E-plane pattern is narrowed down to about 67°, and the H-plane one is larger than 120°. E- and H-plane HPBW ratio, as well as the front-to-back ratio, can be finely controlled by the metallic reflector's shape and size (Noghanian and Shafai 1998, Wong and Luk 2005, Ha et al. 2019).

Figure 2.24 shows the simulated and measured bore-sight gains, which range from 5.8 to 7.5 dBi, with a stable in-band average gain of 6.6 dBi and fluctuation less than 1.7 dB. Therefore, the antenna's available radiation bandwidth can be determined as its impedance bandwidth of 56.2%.

As compared to (An et al. 2012, Chen et al. 2017, Wen et al. 2017, Wong et al 2008, Ge and Luk 2013, Jiang and Werner 2015, Hu et al. 2019, Ding et al. 2017, Ding et al. 2020, Sun et al. 2017) in (Zhao et al. 2021), the 2-D, dual-mode resonant, full-wavelength sectorial dipole can exhibit a series of merits, e.g., wideband, wide beamwidth, less frequency dispersive, and simple configuration. To fully demonstrate its potential for future applications, a dual-polarized, cross-dipole antenna has been implemented and studied in Zhao et al. (2021). Herein, we will show how the dual-polarized antenna can be employed to design a four-element array, with

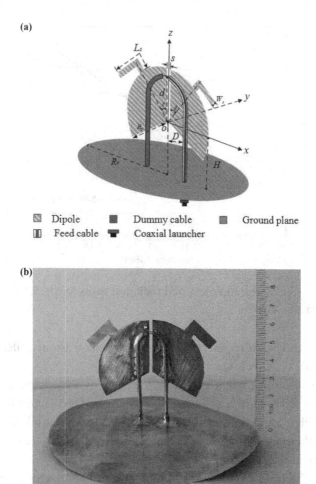

FIGURE 2.21   2-D, Dual-mode resonant dipole antenna mounted above metallic reflector: (a) configuration and (b) photograph of the fabricated prototype.

TABLE 2.3   Parameters of the Sectorial Dipole backed by a Circular Metallic Reflector ($\alpha = 240°$)

	Parameters		
R_0 (mm)	29.1	d(mm)	25
L_S (mm)	20	D (mm)	10
W_S (mm)	5.8	H (mm)	30
β (deg.)	40°	R_r (mm)	60
s (mm)	1.4		

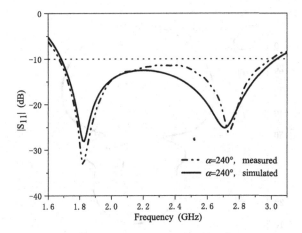

FIGURE 2.22   Simulated and measured reflection coefficients for antenna proto-types mounted above the metallic reflector.

its fabricated prototype shown in Figure 2.25. The elements for each polarization are combined by a four-way, 1:1:1:1 Wilkinson power divider at its basic configuration.

Two-port reflection and isolation coefficients are simulated, measured, and plotted in Figure 2.26. Compared to the dual-polarized element in Zhao et al. (2021), the impedance bandwidth has been degraded to 33% (1.77 to 2.47 GHz) by the narrowband power divider, and the port isolation has been degraded to 20 dB. Hence, the performance is expected to get improved by using an optimally designed feed network.

The normalized simulated and measured radiation patterns at the two ports are shown in Figure 2.27. All of the measured radiation patterns match well with the simulated ones. It is seen the E_φ and E_θ components are nearly equal in both planes and ports, which implies the antenna can exhibit a dual-polarized characteristic. The E_φ and E_θ patterns can be transformed into the co-polarization and cross-polarization ones defined in (Ludwig 1973). It is noted that the dual-polarized patterns presented in Zhao et al. (2021) should be the "co-polarization" and "cross-polarization", rather than the E_φ and E_θ ones. Therefore, the antenna array still exhibits acceptable polarization purity as the element behaves.

The simulated and measured bore-sight radiation gains at two ports are plotted in Figure 2.28. As can be seen, the antenna array can exhibit an average gain of 13 dBi at both ports, which is similar to conventional arrays.

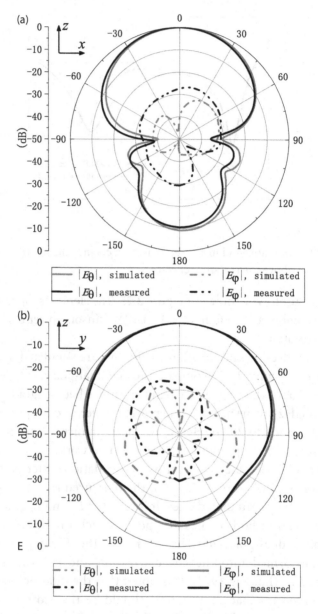

FIGURE 2.23   Simulated and measured normalized radiation patterns: (a) and (b) *xz*- and *zy*-planes at 1.8 GHz;

(*Continued*)

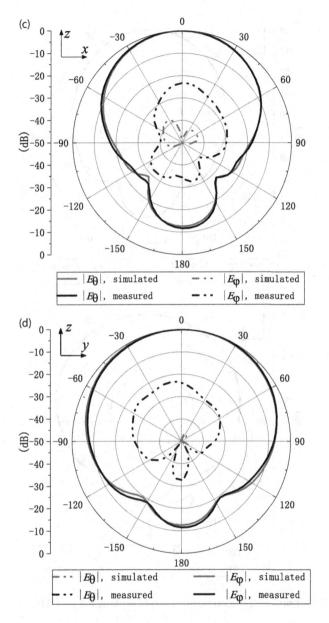

FIGURE 2.23 (*Continued*)   (c) and (d) *xz*- and *zy*-planes at 2.3 GHz;

(*Continued*)

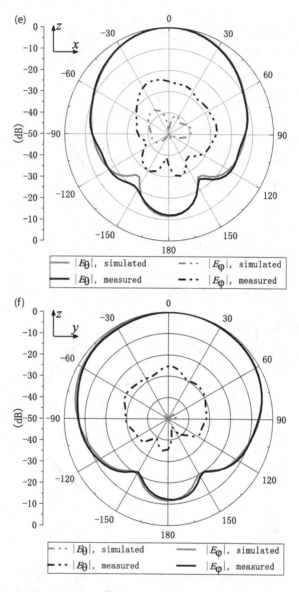

FIGURE 2.23 (*Continued*)   (e) and (f) *xz-* and *zy*-planes at 2.7 GHz.

The in-band fluctuation for both ports is less than 2 dB. As is expected, an improved, optimized feed network would be beneficial to design a high-performance array. The preliminary studies on four-element arrays validate that even a feed network with poor performance is used, the sectorial

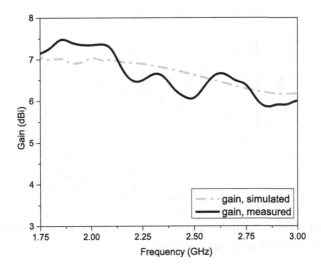

FIGURE 2.24 Simulated and measured radiation gains for antenna prototypes with metallic reflector.

FIGURE 2.25 Photograph of the dual-polarized, four-element antenna array prototype.

dipole array can exhibit dual-mode resonant, wideband characteristics as its element behaves.

By comparing the 1-D and 2-D multi-mode resonant dipole design approach, it is seen the 2-D one can offer one additional degree of freedom

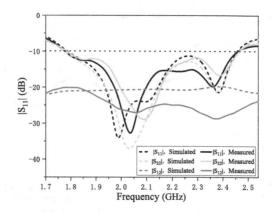

FIGURE 2.26 Simulated and measured reflection coefficients and port isolation for the dual-polarized antenna prototype.

(i.e., the flared angle α) in design. It leads to the mode gauged functionality upon using a gauged electric dipole, which is complementary to the magnetic cases in (Lu et al. 2018) and (Lu et al. 2019). Unlike the traditional 1-D, straight full-wavelength dipoles, the 2-D sectorial full-wavelength dipole can be easily implemented by exciting the fundamental TE mode and its third-order counterpart for broadband operation. It can be directly fed a coaxial cable without incorporating external impedance matching or reactance compensation networks, and other accessories for radiation performance enhancement. As can be seen, the 2-D dual-mode resonant, sectorial full-wavelength dipole antenna exhibits excellent merits such as wideband impedance matching, low-frequency dispersive gain, good polarization purity, and simple configuration. Therefore, it would be a promising candidate for base station antenna in future wireless communications.

More recently, the 2-D dual-mode resonant, sectorial full-wavelength dipole has been employed to design 2-D, multi-mode resonant Yagi–Uda antennas Jia et al. (2021). Unlike conventional Yagi–Uda antennas using linear half-wavelength dipoles, sectorial, full-wavelength dipoles can yield novel designs with smaller separation, fewer elements, and higher gain: When the separation of principal radiator and director is set to 0.024-wavelength, a triple-element Yagi–Uda antenna with gain up to 10.7 dBi can be yielded, with fractional available radiation bandwidth up to nearly 40%. More details can be referenced to Jia et al. (2021).

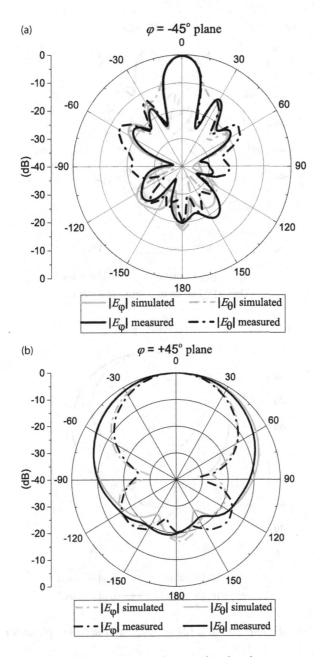

FIGURE 2.27 Simulated and measured normalized radiation patterns for different ports of the dual-polarized antenna at different frequencies: (a) and (b) $\varphi=-45°$ and $\varphi=+45°$ plane for port1 at 2.04 GHz;

(*Continued*)

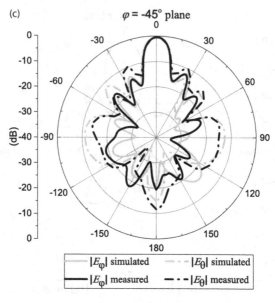

(c) $\varphi = -45°$ plane

$|E_\varphi|$ simulated	$|E_\theta|$ simulated
$|E_\varphi|$ measured	$|E_\theta|$ measured

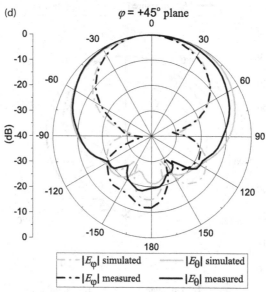

(d) $\varphi = +45°$ plane

$|E_\varphi|$ simulated	$|E_\theta|$ simulated
$|E_\varphi|$ measured	$|E_\theta|$ measured

FIGURE 2.27 (*Continued*) (c) and (d) $\varphi=-45°$ and $\varphi=+45°$ plane for port1 at 2.38 GHz;

(*Continued*)

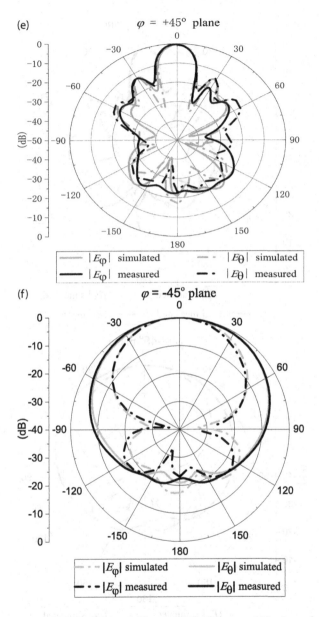

FIGURE 2.27 (*Continued*) (e) and (f) $\varphi=+45°$ and $\varphi=-45°$ plane for port2 at 2.04 GHz;

(*Continued*)

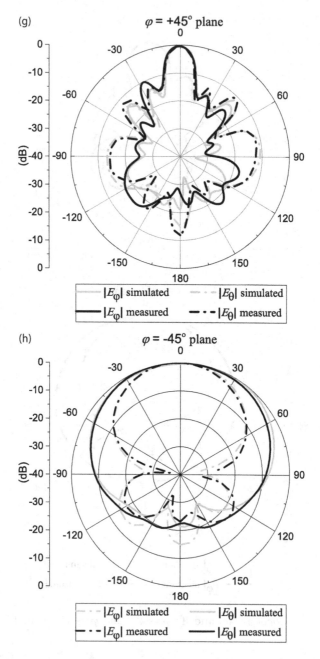

FIGURE 2.27 (*Continued*) (g) and (h) $\varphi=+45°$ and $\varphi=-45°$ plane for port2 at 2.38 GHz.

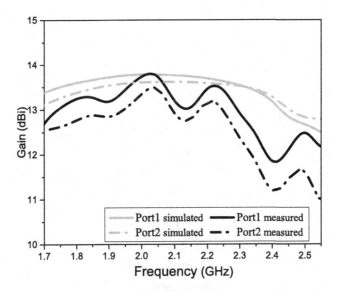

FIGURE 2.28 Simulated and measured radiation gains at different ports of the dual-polarized antenna array.

2.5 VARIANTS OF MULTI-MODE RESONANT ELECTRIC DIPOLES

2.5.1 Symmetric Multi-Mode Resonant, Dual-Band Bent Dipole

If more degrees of freedom in design can be introduced to multi-mode resonant electric dipoles at their basic forms, i.e., a tapered/bent dipole (Lui et al. 2007a, Lu et al. 2012) or an offset-fed bent dipole (Xu et al. 2015), both odd-order and partially excited even-order modes (Xu et al. 2018) can be employed to broaden the operation bandwidth and simultaneously realize dual-band characteristics. In addition, hybrid resonant modes can also be introduced and led to wideband designs.

The configuration of the bent dipole (Lu et al. 2012) is shown in Figure 2.29a, and the fabricated prototype with balancing device (balun), is shown in Figure 2.29b. The bent dipole is composed of a short, symmetric dipole and a long, flared-out, antipodal slot radiator. It is fabricated on a dielectric substrate with relative permittivity of 2.2 and thickness of 1 mm, with most parameters given in (Lu et al. 2012). In this case, the key parameters have been slightly modified as $L_d = 9.2$ mm, and $L_1 + L_t = 29.5$ mm.

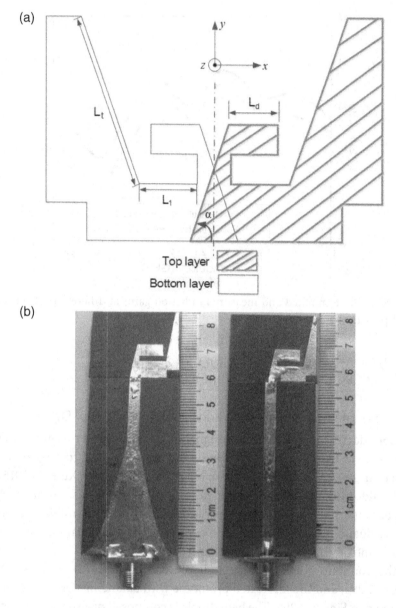

FIGURE 2.29   Symmetric bent dipole under multi-mode resonance: (a) configuration and (b) photograph of the fabricated bent dipole with balun.

Unlike conventional single-ended fed printed antennas (Lui et al. 2007), the balanced, symmetric bent dipole can be readily integrated with other symmetric or differential microwave circuits in direct.

According to Hertz's edge effect Schantz (2015), the lengths of dipole and flared slot portion have been employed to yield a set of approximate design formulas, to outline the antenna's basic radiation behavior, as well as to empirically estimate its initial values of key design parameters (Lu et al. 2012). Figure 2.30 clearly illustrates the possible resonant modes for radiations. We can analyze the radiation behavior from the current distributions, with reference to the coordinate system in (Lu et al. 2012). At 2.4 GHz, the "bent dipole" mode would be the dominant one, and its resonant frequency can be estimated by the sum of tapered, antipodal slot edge and short dipole lengths. In this case, a nearly omnidirectional radiation pattern would be yielded. When the antenna is operating at 5.3 GHz, the short dipole becomes resonant and it may generate directional radiation patterns like a dipole mounted above a metallic reflector. With the increase of operating frequency, high order "bent dipole" and "short dipole" modes would be gradually excited, which may lead to the superposition of hybrid, high-order resonant modes. It yields a more directive, $+y$-oriented, endfire radiation pattern. Since the antenna is center-fed, only the symmetric, odd-order dipole modes can be sufficiently excited,

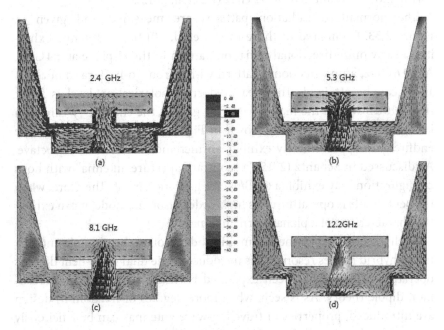

FIGURE 2.30   Simulated surface current distributions at four different frequencies: (a) 2.4 GHz, (b) 5.3 GHz, (c) 8.1 GHz, and (d) 12.2 GHz.

with all anti-symmetric, even-order ones fully suppressed. Therefore, side lobes can be observed at high-frequency bands when the third-order, resonant dipole modes have been excited (Lu et al. 2012).

Another interesting property of the bent dipole should be its angle-dependent input impedance characteristic. Since the bent dipole employs a linearly tapered configuration at its input terminals to enhance its bandwidth, the property of traveling-wave, bi-conical antennas can be incorporated. As illustrated in Figure 2.31, the antenna's input impedance is sensitive to the flared angle α at the input terminal. To attain a less fluctuated resistance/reactance close to 50/0 Ω, α should be set to 80° (Lu et al. 2012). Thus the "bent dipole" can be treated as the combination of truncated traveling-wave and resonant dipoles: It operates as a resonant dipole at a low-frequency band and behaves more likely as a traveling-wave one at a high-frequency band.

The bent dipole antenna is fed by a microstrip to broadside-coupled stripline balun for experimental validations. Figure 2.32 shows the simulated and measured reflection coefficients, which indicates the antenna should exhibit two impedance bandwidths for $|S_{11}|$ lower than −10 dB of 2.41 to 2.61 GHz, and 5.03 to 12 GHz (Lu et al. 2012).

The normalized radiation patterns are measured and given in Figure 2.33. Compared to those in (Lu et al. 2012), the antenna exhibits a nearly omnidirectional radiation pattern in the H-plane at 2.4 GHz. At 10.6 GHz, a unidirectional pattern with a front-to-back ratio of about 10 dB can be attained. The measured cross-polarization level is lower than −10 dB within the main beam. The gain in the endfire, $+y$-direction is simulated, measured, and shown in Figure 2.34. As can be seen, the endfire gain approximately exhibits an increment of about 5 dB/octave. As discussed in Schantz (2015), a "constant aperture antenna" with horn configuration may exhibit a 6 dB/octave gain increment. Therefore, when the bent dipole is operating at its high-order resonant mode, it can exhibit the characteristic of a planar horn antenna.

In this section, a symmetric, multi-mode resonant bent dipole antenna under hybrid modes resonance is introduced. The dual-band bent dipole's operating principle has been explained by using the multi-mode resonant dipole theory. As is seen, when more degrees of freedom in design are introduced, properties of traveling-wave antennas can be felicitously incorporated into a multi-mode resonant dipole, to yield novel designs with distinctive characteristics.

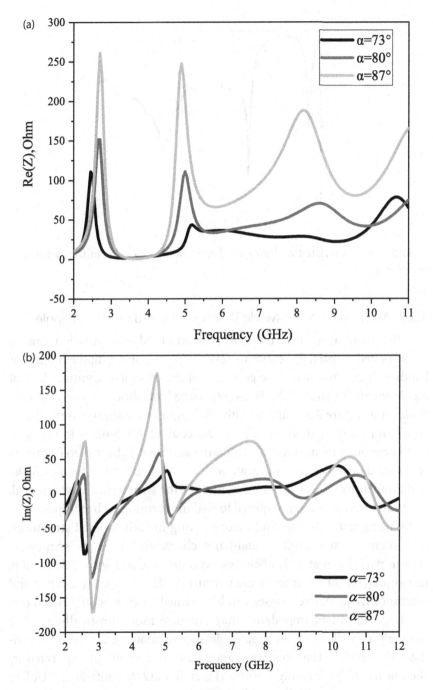

FIGURE 2.31 Simulated input impedance of the symmetric bent dipole: (a) input resistance and (b) input reactance.

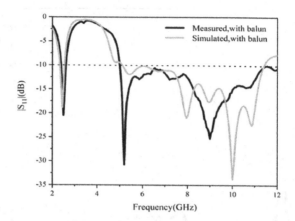

FIGURE 2.32   Simulated and measured reflection efficiencies of the symmetric bent dipole.

2.5.2 Asymmetric Multi-Mode Resonant, Dual-Band Bent Dipole

To offer more available bandwidth, an offset-fed configuration can be incorporated to partially excite the even-order, resonant dipole modes for bandwidth enhancement. The geometry of the offset-fed, asymmetric bent dipole antennas (Xu et al. 2015) and photograph of fabricated prototypes are depicted in Figure 2.35a and b, with the length of the shorter vertical arm (L_2) 0.5 mm shorter than that case in Xu et al. (2015). Similar to the symmetric counterpart in (Lu et al. 2012), the asymmetric bent dipole antenna is designed on a dielectric substrate with $\varepsilon_r=2.2$, loss tangent $\tan\delta=0.0008$, and thinner thickness $h=0.787$ mm. Just like the symmetrical case (Lu et al. 2012), an external balun is required to feed the asymmetric bent dipole.

By using fatter dipoles and incorporating linearly tapered structures, the antenna can exhibit a dual-band characteristic, as its symmetric counterpart (Lu et al. 2012) behaves. As illustrated in Figure 2.36a and b, just similar to the symmetric case, relatively flatten input resistance and reactance frequency responses can be attained when α is 80°. When α is too large, the input impedance may fluctuate more drastically, while a too mall α may lead to low input resistance of about 30 Ω. This well validates the residual traveling-wave antenna characteristic incorporated by the linearly tapered configurations (Lu et al. 2012, Schantz 2015). Unlike the symmetric case in (Lu et al. 2012), the input impedance at the low-frequency band is less sensitive to α.

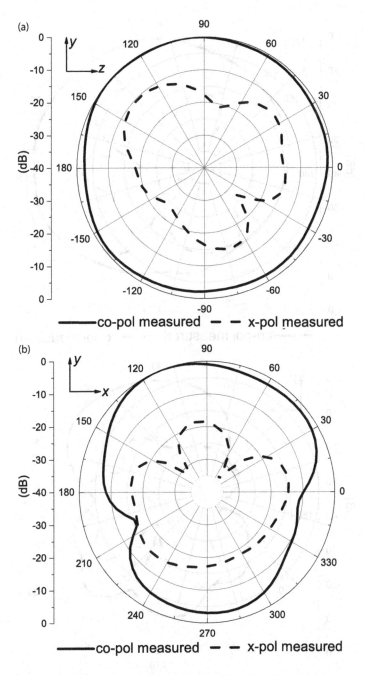

FIGURE 2.33 Simulated and measured normalized radiation patterns of the symmetric bent dipole: (a) H-plane, 2.4 GHz, (b) E-plane, 2.4 GHz,

(*Continued*)

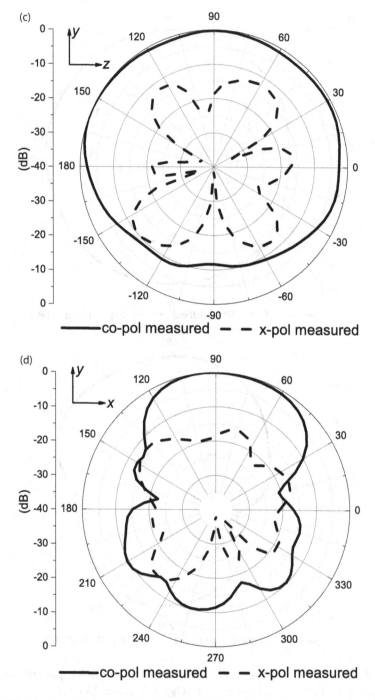

FIGURE 2.33 (*Continued*) (c) H-plane, 10.6 GHz, and (d) E-plane, 10.6 GHz.

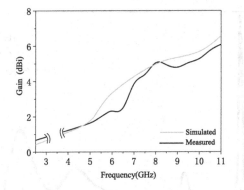

FIGURE 2.34 Simulated and measured symmetric bent dipole gains in the end-fire, $+y$-direction.

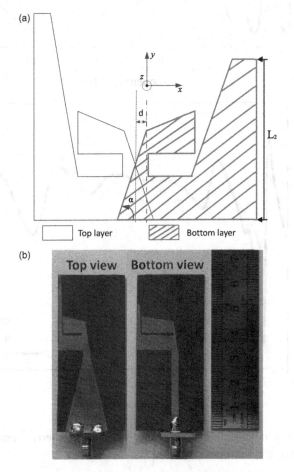

FIGURE 2.35 Asymmetric bent dipole: (a) configuration and (b) photograph of some fabricated prototypes with balun.

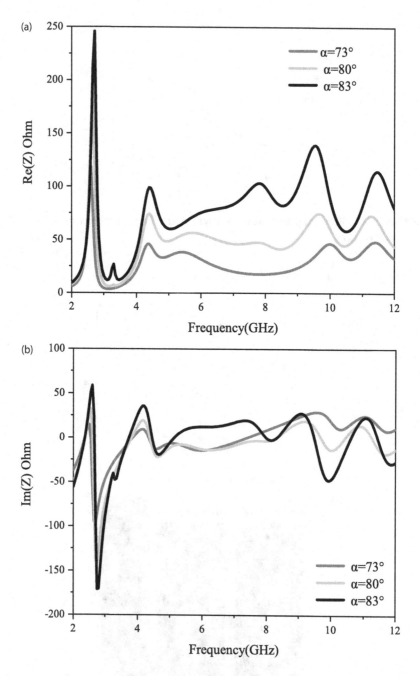

FIGURE 2.36 Simulated input impedance of the asymmetric bent dipole: (a) input resistance and (b) input reactance.

Since the antenna is asymmetric, while it is fed by a balun, it is necessary to study the balun's effect at first. As illustrated in Figure 2.37, when the antenna is simulated without a balun, six resonances can be observed. When it is connected to a balun, nine resonances would be attained. This phenomenon indicates that the reflection coefficient of asymmetric bent dipole should be more sensitive to the balun than its symmetric counterpart behaves (Lu et al. 2012). Nevertheless, both upper and lower bounds of the impedance bandwidths for $|S_{11}|$ lower than $-10\,$dB is barely changed, which implies that an external balun can be used to feed the antenna, just like the symmetric case.

Figure 2.38 shows the simulated and measured reflection coefficients, which indicates the antenna should exhibit impedance bandwidths for $|S_{11}|$ lower than $-10\,$dB of 2.38 to 2.62 GHz at the first band. Compared to the symmetric case, the lowest cut-off frequency of the second band has been effectively reduced to 4.12 GHz, thus it yields an impedance bandwidth from 4.12 to 15.56 GHz, which is comparable to that in (Xu et al. 2015). In this case, it has been evidently proved that the even-order resonant mode

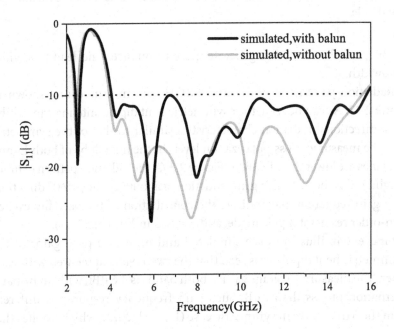

FIGURE 2.37 Effect of balun on the reflection coefficient of an asymmetric bent dipole.

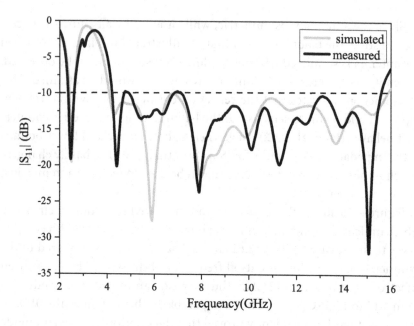

FIGURE 2.38 Simulated and measured reflection coefficients of the asymmetric bent dipole.

can be partially excited to generate more resonances and lead to a wider bandwidth.

Radiation patterns at 6 GHz and 9.8 GHz are measured and shown in Figure 2.39. As can be seen, the asymmetric bent dipole antenna can exhibit quasi-directional radiation characteristics similar to that of the symmetric one. The measured cross-polarization levels in the main lobe of both E- and H- planes are lower than 15 dB. In Figure 2.39b, the radiation pattern exhibits a slightly tilted beam with a gain reduction of about 5 dB in $\varphi=60°$ direction. Such gain reduction may indicate the contribution of the partially excited even-order resonant dipole mode, as discussed in Xu et al. (2015).

Figure 2.40 illustrates the simulated and measured peak gains of the asymmetric bent dipole. It is seen that the two results agree well with each other. The measured average gain at both bands is 4.6 dBi, with an in-band fluctuation of less than 3 dB. Such gain frequency response is different from the 5 dB/octave in symmetric case (Lu et al. 2012), which implies that an asymmetric configuration may be effective to yield a "constant gain antenna" Schantz (2005). Hence the asymmetric bent dipole can achieve a flatten gain frequency response and gain bandwidth. Since the radiation

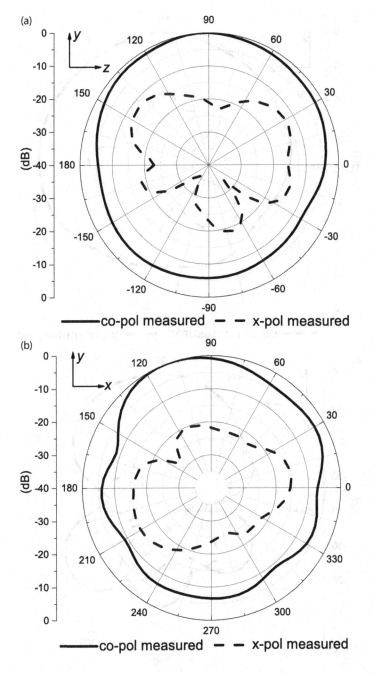

FIGURE 2.39 Measured normalized radiation patterns of the asymmetric bent dipole: (a) H-plane, 6 GHz, (b) E-plane, 6 GHz,

(*Continued*)

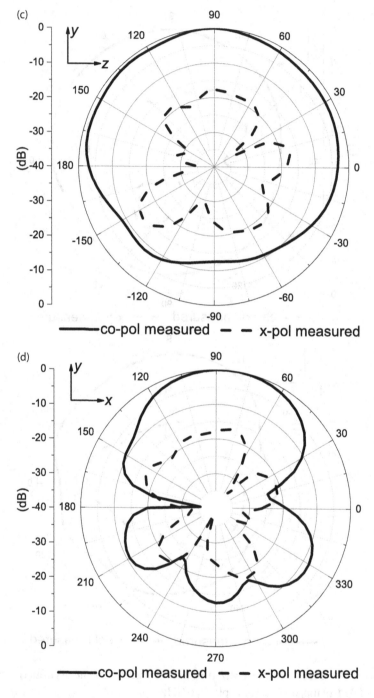

FIGURE 2.39 (*Continued*)   (c) H-plane, 9.8 GHz, and (d) E-plane, 9.8 GHz.

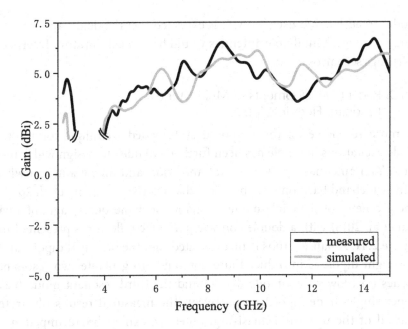

FIGURE 2.40   Simulated and measured peak gain of the asymmetric bent dipole.

patterns rarely change all over the high-frequency band from 4.12 GHz to nearly 15.6 GHz (Xu et al. 2015), the asymmetric bent dipole can not only exhibits a wide impedance bandwidth, but also the identical radiation pattern bandwidth, and gain bandwidth up to 4:1.

In Sections 2.5.1 and 2.5.2, multi-mode resonant bent dipoles with symmetric and asymmetric configurations have been studied in detail. It is found that both antennas can exhibit dual-band, multi-mode resonant characteristics. An off-center feeding may offer wider bandwidth at the high-frequency band by partially exciting the even-order dipole mode, at the cost of slightly tilted radiation patterns at the corresponding band. The beam tilting characteristic compensates for the 5 dB/octave gain increment of horn configuration Schantz (2015), and it yields a relatively flattened gain frequency response. Both antennas can exhibit nearly omnidirectional radiation patterns at the low-frequency band and a unidirectional one at the high-frequency band. Such property is unique and possibly useful in wireless communications: At the low-frequency band, the antennas can operate at "omnidirectional, low data-rate, broadcasting mode" to open, unauthorized users. At the high-frequency band, the antenna can operate at "unidirectional, high data-rate communication mode" to authorized users only. It is expected that the dual-band, multi-mode resonant bent dipole antennas would be applied in future Internet-of-Things scenarios.

2.5.3 Recent Developments of Multi-Mode Resonant Electric Dipoles

In more recent years, the proposed stub-loaded technique for multi-mode excitations in dipole has been further extended to design wideband quasi-Yagi antennas Xu et al. (2018) and wideband dipole antenna with ultra-wideband harmonic suppression characteristics (Zhang et al. 2019). The geometry of the triple-mode resonant wideband quasi-Yagi antenna (Xu et al. 2018) with a double concave parabolic reflector is presented in Figure 2.41. A pair of stubs is incorporated into the principal dipole, and a straight dipole is introduced and combined to generate two resonant modes (i.e., lower dual-mode dipole) and the third resonant mode (i.e., upper single-mode dipole), respectively. The measured results illustrate that all of the wideband quasi-Yagi antennas exhibit broad impedance bandwidths, good unidirectional radiations, high radiation efficiencies, and low cross-polarization levels within the operating frequency bands.

The configuration of the wideband dipole antenna (Zang et al. 2019) with ultra-wideband harmonic suppression and enhanced bandwidth is depicted in Figure 2.42. Herein, the "harmonic" should refer to the "high

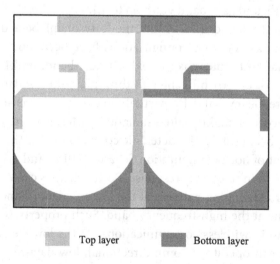

Top layer   Bottom layer

FIGURE 2.41 Configuration of the printed, triple-mode resonant quasi-Yagi antenna (Xu et al. 2018).

order resonant mode" in the opinion of this book. U-shaped patch, slender stub, and parasitic patch have been applied to broaden the working bandwidth of dipole antenna, as well as to improve the bandwidth with high order mode suppression. As is reported, the antenna has a fractional bandwidth of 42.8% (from 3.43 to 5.3 GHz) and a wide harmonic suppression bandwidth ranging from 6.2 to 26 GHz. The high order mode suppression characteristic may be beneficial for mutual coupling reduction and multiple-antenna coexisting in engineering (Zang et al. 2019).

Meanwhile, more variants of multi-mode resonant dipoles (Wen et al. 2017, Hu et al. 2019, Luo et al. 2019, Zhang et al. 2020, Li and Li 2020) have also been presented. The geometry of the wideband antenna (Wen et al. 2017) with a stable omnidirectional radiation pattern is presented. It can be treated as the combination of a dipole and an adjacently coupled loop antenna, as illustrated in Figure 2.43. A wide impedance bandwidth of 44.2% and a stable radiation pattern at both E-plane and H-plane are obtained because of the simultaneous excitation of the antenna's first two resonant modes that share a similar omnidirectional radiation pattern.

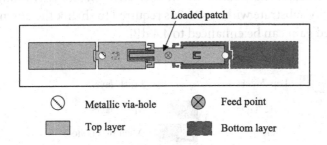

FIGURE 2.42 Configuration of the wideband dipole antenna with high-order resonant mode suppression functionality (Zang et al. 2019).

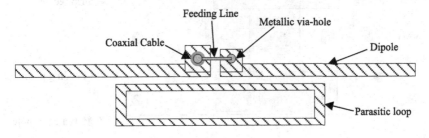

FIGURE 2.43 Geometry of the wideband dipole antenna loaded by loop resonator (Wen et al. 2017).

The configuration of the wideband, multi-layered folded dipole antenna (Hu et al. 2019) with multi-mode resonant modes is shown in Figure 2.44. Three resonant modes are obtained by using a modified planar folded dipole and its coupled feeding structure. Furthermore, multiple resonant modes in the antenna are manipulated, tuned, and then combined to increase the impedance bandwidth by incorporating the shorting pins and parasitic patches. A bandwidth of 80% and a good omnidirectional radiation performance are achieved. Furthermore, the antenna exhibits a flat gain variation of less than 1.3 dB in the horizontal plane.

The configuration of the bandwidth-enhanced, mode compression dipole antenna (Luo et al. 2019) with the two stubs is shown in Figure 2.45. By introducing the loading stubs, the fifth-order resonant mode of the compressed dipole can be tuned close to the third-order one, yielding a wideband operation with enhanced gain. The measurement shows that the gain of the compressed dipole is over 4 dBi, which is higher than that of a conventional dipole over the desired operating band of 3.3 to 3.6 GHz for the fifth-generation mobile communication applications. The mode compression effect strongly relies on the dielectric substrate. Although high permittivity substrate with $\varepsilon_r = 11.2$ is required to shrink the antenna size, its realized gain can be enhanced to 4.4 dBi.

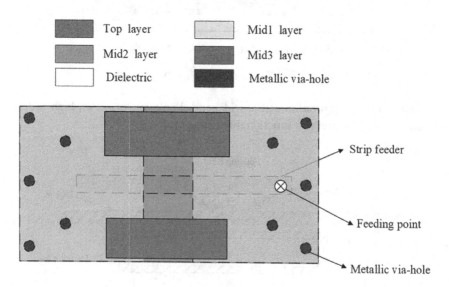

FIGURE 2.44 Configuration of the wideband, multi-layered folded dipole antenna with multiple resonant modes (Hu et al. 2019).

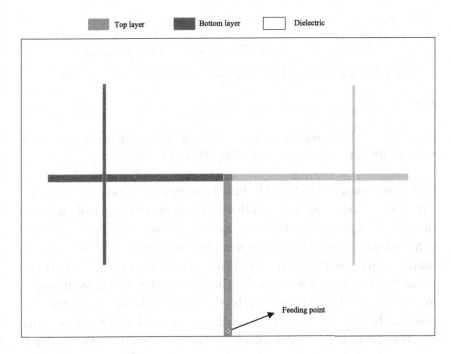

FIGURE 2.45 Configuration of the bandwidth-enhanced mode compression dipole antenna (Luo et al. 2019).

The configuration of the dual-mode resonant dipole antenna Zhang et al. (2020) with an electrically small loop (ESL) resonator is presented in Figure 2.46. An ESL is properly loaded to compress the operating frequencies of first- and third-order modes. Due to the existence of ESL, and the fact that ESL should be equivalent to an orthogonal magnetic dipole, the

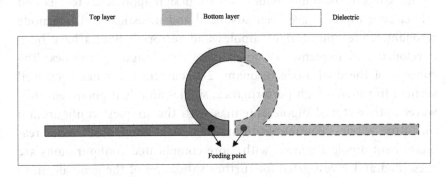

FIGURE 2.46 Configuration of the dual-mode dipole antenna using an open-loop resonator (Zhang et al. 2020).

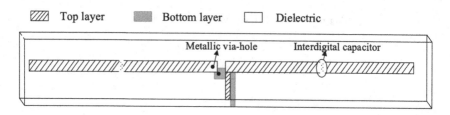

FIGURE 2.47 Configuration of the wideband, dual-mode resonant dipole antenna with dual-point capacitive loadings (Li and Li 2020).

third-order resonant mode can be tuned downward its first-order counterpart. This leads to a wide bandwidth with stable polarization and radiation patterns. The configuration of a wideband dipole antenna (Li and Li 2020) with dual-point capacitive loadings is shown in Figure 2.47. Two capacitors (lumped or inter-digital) are loaded near the current distribution nulls of the third-order resonant mode. Complementary to the strip-loaded slotline case in (Lu and Zhu 2015b), the resonant frequency of the first mode has been tuned upward the third one, with little effects on the latter one. Thus it yields a compression of two resonant modes and a wideband property. A good performance including an effective bandwidth of 51.4% (from 1.88 to 3.18 GHz) and an omnidirectional radiation pattern with gain varying from 2.0 to 4.0 dBi in an azimuthal plane is obtained within a thin size of $1.01 \times 0.09\,\lambda_0$. Compared to existing design approaches, the capacitive-loaded design approach can maintain the inherent slim profile of resonant electric dipole without incorporating cross stubs.

2.6 CONCLUDING REMARKS

In this chapter, the multi-mode resonant design approaches to 1-D and 2-D electric dipoles have been advanced and investigated. Dual-mode resonant, wideband straight dipoles and sectorial dipoles have been developed and implemented to validate the design approaches. The concept of the dual-mode resonant, 2-D current sheet has been well verified to provide high-performance, single- and dual-polarized, full-wavelength sectorial dipole antennas with the simplest configuration for future wireless communications. Then, dual-band, multi-mode resonant bent dipole antennas with more complicated configurations are presented and investigated for further validation of the generalization of the multi-mode resonant design approach and concepts. Finally, relevant variants of multi-mode resonant dipoles have been introduced and discussed. Besides wideband operation, high-order resonant modes can also be employed to enhance the directivity and gain of electric and magnetic dipole antennas (Jennetti, and Uda 1967, Landsdorfer 1976, Lu et al. 2019, Luo et al. 2019). More relevant examples for applications will be raised and discussed in Chapter 6.

CHAPTER 3

Multi-Mode Resonant Slot and Loop Antennas

3.1 BRIEF HISTORY AND RECENT DEVELOPMENTS OF SLOT ANTENNAS

The conceptual design of slot antennas can be dated back to the late 1930s (Blumlein 1939). A slit, tubular conductor excited by coil should be the first prototype of a slot antenna at a high-frequency band. As is commonly treated as a "linear magnetic dipole", the width of a "slot antenna" should be much smaller than its resonant length and operating wavelength, and the ground plane's area around the aperture should approach infinity. In this manner, the antenna should employ the virtual surface magnetic currents distributed within the slim aperture to contribute radiations. Under such definition, and is recognized as the complementary counterpart of electric dipole (Bailey 1946, Booker 1946), slot antenna has been widely studied in the past few decades since the late 1940s, with its developments briefly depicted in Figure 3.1a and b. Since the 1950s, traveling-wave slot antennas and resonant slot antennas have been mathematically modeled and analyzed for their radiation behaviors and used in many flush-mounted, aviation applications, as illustrated in Figure 3.1c–e Stephenson and Walter (1955), Oliner (1957a, b), Johnson (1955), Hsieh et al. (1971).

Besides the bulky, traveling-wave, tapered slot antennas (Hines et al. 1953, Gibson 1973, Lai 1992, Shin and Schaubert 1999), resonant slot antennas generally exhibit relatively compact size but narrow bandwidth

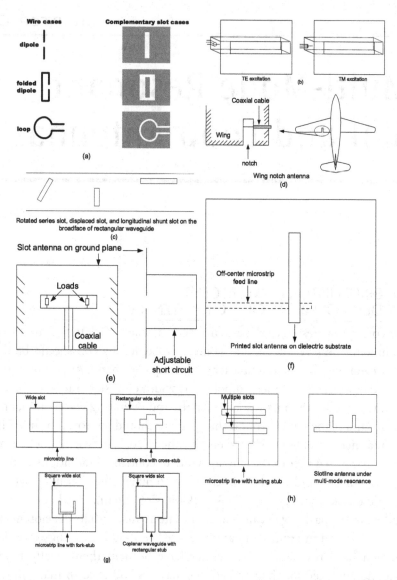

FIGURE 3.1   Brief history of slot antennas: (a) slot antennas/elements and their corresponding complementary wire counterparts (Bailey 1946), (b) traveling-wave slot antenna under different excitations (Booker 1946), (c) waveguide slot antennas (Stephenson and Walter 1955, Oliner 1957a,b), (d) wing notch antenna (Johnson 1955), (e) slot antenna with end loading Hsieh et al. (1971), (f) off-center fed microstrip slot antenna (Yoshimura 1972), (g) wide-aperture slot antenna (Kahrizi et al. 1993, Jang 2000, Sze and Wong 2001, Chen 2003), and (h) multi-slot antenna (Lui et al. 2005) and multi-mode resonant slotline antenna (Lu and Zhu 2015a, b).

due to their inherent configuration. To attain wideband operation, various design approaches have been developed since the 1970s: An offset microstrip-fed technique to narrow microstrip slot antennas has been advanced, as illustrated in Figure 3.1f for wideband design (Yoshimura 1972) at first. Figure 3.1g shows the "wide aperture" configurations excited by single-stub, or multi-digit and patch-like tuning stubs (Kahrizi et al. 1993, Jang 2000, Sze and Wong 2001, Chen 2003). In these designs, the overall ground plane's area around the aperture has been shrunk to comparable with the aperture's size. Then, the reshaped "big aperture" antennas may exhibit wideband operation under hybrid, multiple resonances at the cost of more complicated configuration and unidentifiable modes Pan et al. (2014). To maintain the pure resonant "slot mode" of slot antennas, the inherent narrow slot configuration should be maintained. For broadband designs of narrow slot antennas, external resonators Zhu et al. (2003), reactance compensation networks (Behdad and Sarabandi 2004), or multiple narrow slots Lui et al. (2005) can be employed, at the cost of more complicated configurations with bulky size.

Based on the 1-D multi-mode resonant dipole theory, wideband, shared-aperture antennas with narrow slot configurations (Lu and Zhu 2015a, b, c) have been advanced and realized: Multiple resonant modes within a single slotline radiator can be simultaneously excited without introducing external accessories or additional radiators. This leads to the multi-mode resonant design approach to slotline antennas shown in Figure 3.1h.

3.2 CENTER-FED MULTI-MODE RESONANT SLOTLINE ANTENNAS

The odd-order resonant modes of a slotline antenna fed at its center can be simultaneously excited, with all even-order ones sufficiently suppressed. A series of design approaches have been advanced to the multi-mode resonant slotline antennas' behavior: As graphically illustrated in Figure 3.2a, the 1.5-wavelength resonant mode can be perturbed downward its fundamental counterpart (Lu and Zhu 2015a, c) by incorporating short slits near its E-field distribution nodes. Figure 3.2b shows that short-circuited strips can be used to tune the fundamental mode upward its third-order counterpart (Lu and Zhu 2015b). Accordingly, both fundamental and fifth-order, 2.5-wavelength modes can be perturbed to the 1.5-wavelength one, then to yield a triple-mode resonant slotline antenna Guo et al. (2017), as illustrated in Figure 3.2c. As presented in Lu and Zhu (2015a, b, c),

Guo et al. (2017), the bandwidth of basic slotline antennas can be effectively enhanced to near or over 30% under multiple-mode resonance without externally incorporated accessories.

3.2.1 Dual-Mode Resonant Slotline Antenna with Loaded Stubs

Figure 3.3 depicts the center-fed, stub-loaded slotline antenna having length L and width W (Lu and Zhu 2015a, Lu and Zhu 2015c), with surface E-field distribution magnitudes of both half- and 1.5-wavelength resonant modes shown. In this section, a slotline antenna designed on a substrate with relative permittivity $\varepsilon_r=2.65$, thickness $h=1.0$ mm will be investigated at a 3.5 GHz band.

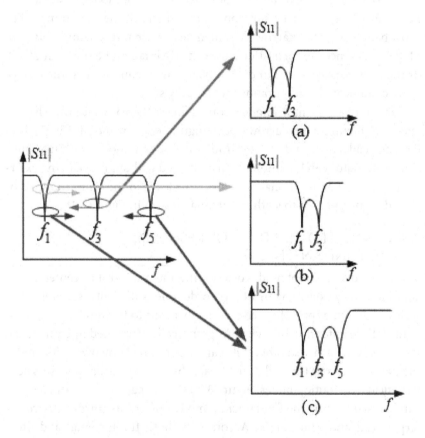

FIGURE 3.2 Conceptual illustration to the $|S_{11}|$ frequency responses of dual- and triple-mode resonant, center-fed slotline antennas using the first three odd-order resonant modes.

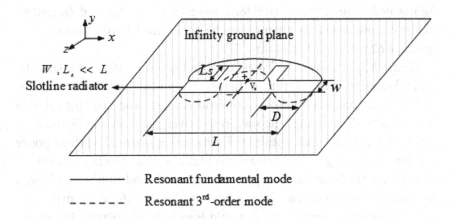

FIGURE 3.3 Center-fed stub-loaded slotline antenna with electric field distributions of the first two, odd-order resonant modes.

A pair of slits with length L_s can be incorporated at the E-field distribution node of 1.5-wavelength mode, for mode excitation and reallocation. Figure 3.4 shows the resonant characteristic of the slotline antenna with a different length of L_s. As can be seen, the 1.5-wavelength mode can be excited and reallocated downward the fundamental one, such that a wideband, dual-mode resonant, reflection coefficient frequency response can

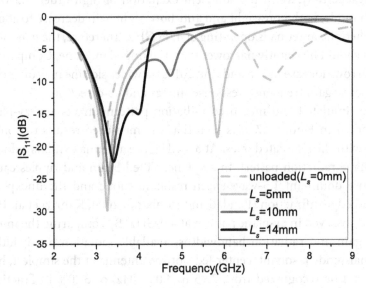

FIGURE 3.4 Dual-resonance reallocation with respect to stub length L_s ($W = 6$ mm, $D = 17$ mm, $L = 62$ mm).

be attained. Using the empirical formulas presented in (Lu and Zhu 2015a, c), the parameters of the antenna can be determined as $W = 6$ mm, $D = 17$ mm, $L = 62$ mm, and $L_s = 14$ mm.

Figure 3.5a–c shows the E-field distributions of uniform, unilaterally and bilaterally loaded slotline radiators. For odd-order resonant slotline radiator, the natural boundary conditions in xz- and zy-planes should be electric wall and magnetic wall, respectively, as shown in Figure 3.5a. Thus, the E-field lines within the slits are x-oriented with opposite directions owning to the natural magnetic wall in zy-plane, as shown in Figure 3.5b. To maintain a pure electric wall condition in xz-plane, a bilaterally loaded case in Figure 3.5c should be beneficial for maintaining the antenna's symmetry: It would lead to high polarization purity than the unilaterally loaded one (Lu and Zhu 2015a, c). A photograph of a fabricated prototype fed by a directly bonded coaxial cable is shown in Figure 3.5d.

Figure 3.6 depicts the simulated and measured reflection coefficient and gain frequency responses of the fabricated slotline antenna. As can be seen, the measured impedance bandwidth for $|S_{11}|$ lower than −10 dB is from 2.95 to 4.15 GHz, which is about 33.8% in fraction. The average bore-sight and anti-bore-sight gains in +z- and −z-directions are 3.5 and 3.0 dBi, respectively. With the sufficient excitation of high-order, 1.5-wavelength mode, both bore-sight and anti-bore-sight gain decrease to about 0 dBi when the antenna is operating at 4.05 GHz. Therefore, the gain bandwidth should be slightly narrower than the impedance one. Compared to the narrow slot case in (Lu and Zhu 2015a), a wider slotline may offer more symmetric gain frequency responses in +z- and −z-directions.

The simulated and measured radiation patterns in xz- and zy-planes are shown in Figure 3.7. It is seen that the measured results reasonably agree with the simulated ones. At 3.4 GHz, the antenna exhibits a dough-nut shape radiation pattern in zy-plane. Tilted beam and ripples caused by more dominant 1.5-wavelength resonant mode and the unexpected ground edge diffraction (Behdad and Sarabandi 2004, Kahrizi et al. 1993) can be observed in zy-plane pattern at 4.0 GHz. By comparing the imped-ance, gain and radiation bandwidths, available radiation bandwidth of the dual-mode resonant, center-fed slotline antenna at the simplest, basic form can be recognized from 2.95 to 4.05 GHz, or 31.0% in Fractional bandwidth (FBW), which is comparable to those presented in Lu and Zhu (2015a and c).

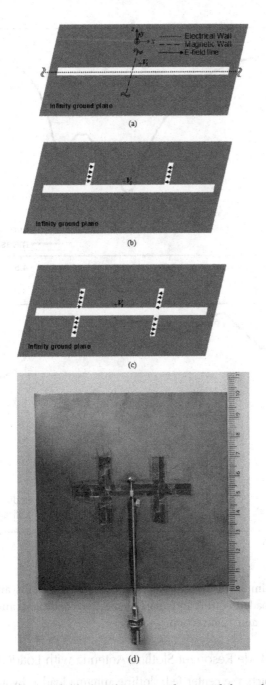

FIGURE 3.5 Surface E-field distribution and natural boundary conditions of slotline antennas: (a) unloaded antenna, (b) and (c) unilateral and bilateral stub-loaded antennas, and (d) photograph of a fabricated prototype.

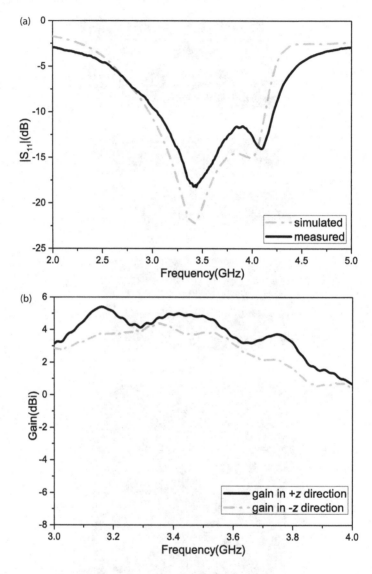

FIGURE 3.6 Simulated and measured reflection coefficients and gain of the bilateral stub-loaded slotline antenna: (a) reflection coefficients and (b) measured gain in both +z- and –z-direction.

3.2.2 Dual-Mode Resonant Slotline Antenna with Loaded Strips

Figure 3.8 depicts the center-fed, slotline antenna loaded by a pair of short-circuited strips (Lu and Zhu 2015b). As can be seen, short-circuited strips are placed near the E-field nodes of the 1.5-wavelength mode so that they can be

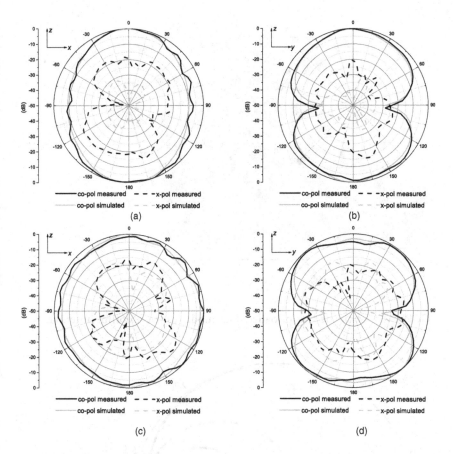

FIGURE 3.7 Simulated and measured radiation patterns: (a) xz-plane, 3.4 GHz, (b) zy-plane, 3.4 GHz, (c) xz-plane, 4.0 GHz, and (d) zy-plane, 4.0 GHz.

excited. Simultaneously, the fundamental one can be tuned upward its third-order counterpart to yield a dual-mode resonant slotline antenna. In this section, the investigated slotline antenna is designed on a substrate with relative permittivity $\varepsilon_r=2.65$ and thickness $h=1.0$ mm, with most parameters identical to those in (Lu and Zhu 2015b), but a larger via-hole diameter $d=1.5$ mm.

Figure 3.9a shows the Smith Chart of the slotline antennas using different loaded strip lengths of L_s. In the unloaded case ($L_s=0$), the impedance trace is less concentrated than the loaded case. When the antenna is loaded by a pair of short-circuit strips, the input impedance trace may converge to the right-hand side of the Smith chart, which yields a multi-mode resonant status with high average input resistance up to 300 to 500 Ω (Lu and Zhu 2015b).

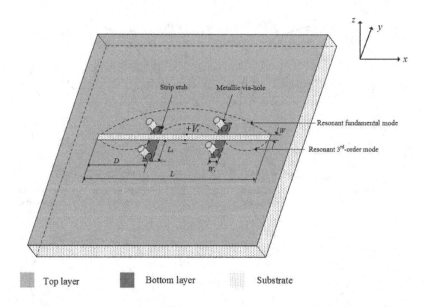

FIGURE 3.8   Center-fed strip-loaded dual-mode resonant slotline antenna.

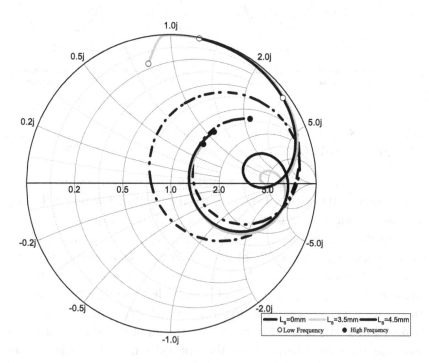

FIGURE 3.9   Impedance of slotline antennas using a different loaded strip length of L_s.

To match with a 50 Ω coaxial cable and the test instruments, a linearly tapered, microstrip impedance transformer is employed to feed the antenna, as shown in Figure 3.10a. A photograph of fabricated prototype is presented in Figure 3.10b. Figure 3.11 plots the simulated and measured reflection coefficients. The measured reflection coefficient exhibits a dual-mode resonant frequency response with impedance bandwidth for $|S_{11}|$ lower than −10 dB from 3.5 to 5.3 GHz, i.e., 40.9% in fraction. Compared

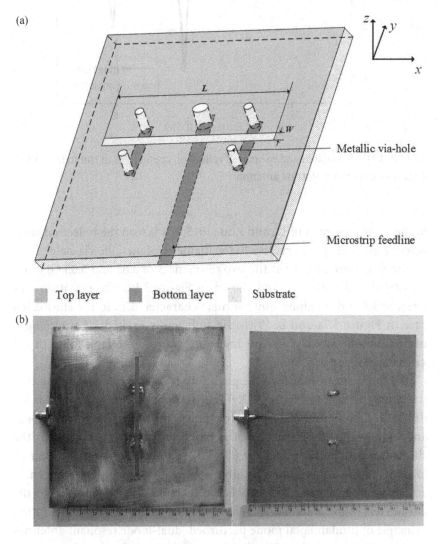

FIGURE 3.10 The strip-loaded, dual-mode resonant slotline antenna: (a) configuration and (b) photograph of fabricated prototypes.

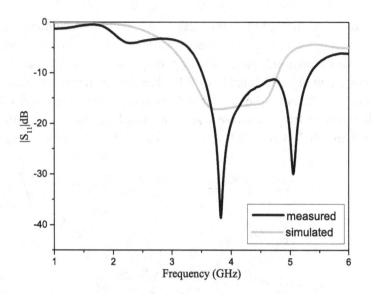

FIGURE 3.11   Simulated and measured reflection coefficients of the strip-loaded, dual-mode resonant slotline antenna.

to the cases presented in Lu and Zhu (2015b), it is seen the reflection coefficient should be quite sensitive to the diameter of metallic via-hole.

The radiation patterns at the two resonances of 3.83 and 5.05 GHz are simulated and measured and shown in Figure 3.12. The radiation patterns at 3.83 GHz exhibit quite similar characteristics to the slit-loaded case in Figure 3.7a and b. With the sufficiently excited 1.5-wavelength resonant mode at a high-frequency band, ripples and triple lobes in the xz-plane can be observed.

The frequency response of the peak gain is measured and compared to the simulated one in Figure 3.13. The measured average peak gain is 5.0 dBi, with a maximum discrepancy of about 2 dB within the whole impedance bandwidth. By comparing the radiation patterns and gain to the slit-loaded case Lu and Zhu et al. (2015a), it can be inferred that the strip-loaded slotline antenna may exhibit a more frequency dispersive characteristic. In addition, the frequency dispersion is quite sensitive to the parameters of the antenna. Therefore, the design approach and operation principle of fundamental mode perturbed, dual-mode resonant antennas are expected to be useful for effectively controlling the radiation dispersion of resonant wideband antennas.

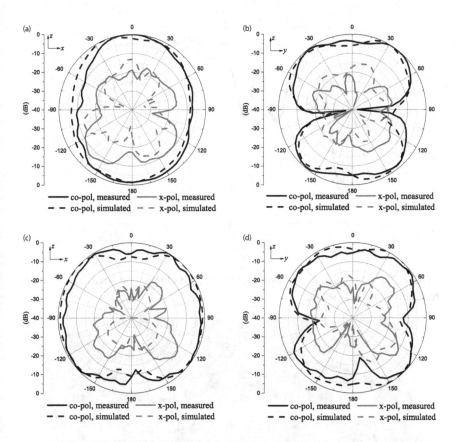

FIGURE 3.12 Simulated and measured radiation patterns: (a) xz-plane at 3.83 GHz, (b) zy-plane at 3.83 GHz, (c) xz-plane at 5.05 GHz, and (d) zy-plane at 5.05 GHz.

3.2.3 Triple-Mode Resonant Slotline Antenna

Figure 3.14 depicts the initial center-fed, triple-mode resonant slotline antenna having a length of 1.5-wavelength Guo et al. (2017). The slotline antenna is designed at 4.0 GHz, on a substrate with relative permittivity $\varepsilon_r = 2.65$ and thickness $h = 1.0$ mm, with most design parameters identical to that in Guo et al. (2017), except the width of fork-digit (W_{h4}) and the diameter of via-hole is set as 0.6 mm, and the slot is 1 mm longer. To simultaneously excite the first three, odd-order resonant modes in Figure 3.14, a microstrip line with stepped-impedance, a short-circuited stub is symmetrically fed to the slotline radiator, as shown in Figure 3.15a. Next, two additional, short-circuited stubs are employed to excite and tune the fundamental mode (Lu and Zhu 2015b), and the stubs are connected

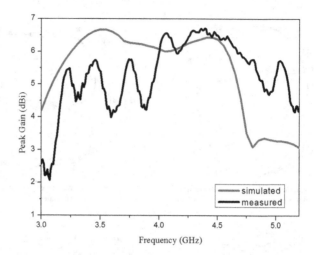

FIGURE 3.13 Simulated and measured peak gain of the strip-loaded dual-mode resonant slotline antenna.

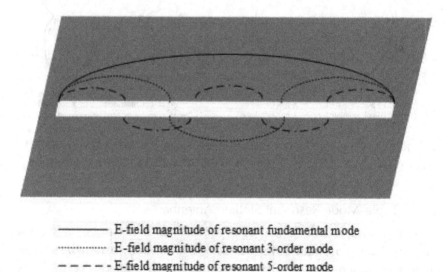

——————— E-field magnitude of resonant fundamental mode

················· E-field magnitude of resonant 3-order mode

— — — — - E-field magnitude of resonant 5-order mode

FIGURE 3.14 Center-fed slotline antenna with electric field distributions of the first three, odd-order resonant modes.

to the center feed line, to form a triple-digit feed network shown in Figure 3.15b. On the other hand, slits can be incorporated into the slotline radiator at the outer E-field node of the fifth-order resonant mode, as illustrated in Figure 3.15c. Therefore, using the combination of the

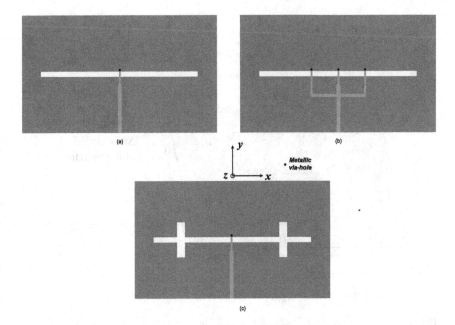

FIGURE 3.15   Evolution of the multiple-mode slotline antennas: (a) center-fed by single short-circuited microstrip stub, (b) center-fed by fork-shape microstrip stubs, and (c) center-fed by single short-circuited microstrip stub and perturbed by slits.

latter two cases in Figure 3.15b and c, the triple-mode resonant slotline antenna in Figure 3.16a can be yielded. The photograph of a fabricated prototype is shown in Figure 3.16b.

Simulated and measured reflection coefficients of the triple-mode resonant slotline antenna are shown in Figure 3.17. A triple-mode resonant frequency response for the reflection coefficient can be intuitively observed and evidently confirmed. The impedance bandwidth for $|S_{11}|$ lower than −10 dB is 3.17 to 4.43 GHz, i.e., 33.1% in fraction. Compared to the case in Guo et al. (2017), the bandwidth is slightly narrower, with the first two resonances reallocated closer.

The xz-plane radiation patterns at 3.42, 3.82, and 4.22 GHz are measured and plotted in Figure 3.18. Only the co-polarized, E_φ component is presented and compared. The antenna can exhibit relatively stable gain in the +z- and −z-directions, owning to the superposition of exited, odd-order resonant modes. Interestingly, the antenna can symmetrically exhibit frequency scannable radiation nulls within a narrow angle range from about $\theta=\pm60°$ to $\theta=\pm20°$. Such a phenomenon indicates that the antenna may

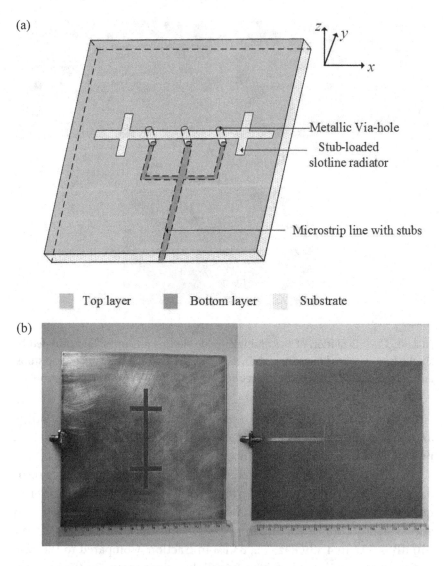

(a)

Metallic Via-hole

Stub-loaded
slotline radiator

Microstrip line with stubs

Top layer   Bottom layer   Substrate

(b)

FIGURE 3.16   Triple-mode resonant slotline antenna: (a) configuration and (b)
photo of fabricated prototypes.

exhibit a null frequency scanning functionality. Conventionally, such dispersive radiation characteristics can only realize by arrays in the single-
mode resonant case, with the null(s) steered by the array factor.

Figure 3.19 shows the simulated and measured peak gain of the
triple-mode resonant slotline antenna. Unlike the fundamental mode
perturbed-only case in Lu et al. (2015b), the triple-mode resonant

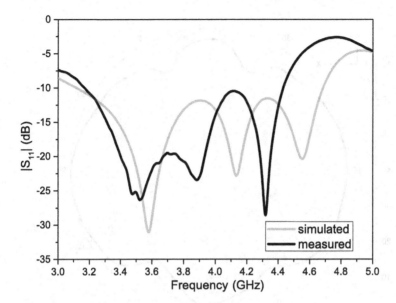

FIGURE 3.17   Simulated and measured reflection coefficients of the triple-mode resonant slotline antenna.

slotline antenna's gain frequency response sounds less frequency dispersive and behaves quite similar to that of the slit-loaded case in Lu et al. (2015a). Heuristically, as can be seen from the measured radiation pattern and gain, the radiation dispersion characteristic of multi-mode resonant antennas can be effectively controlled by the usable resonant modes. The revealed operation principles of 1-D, multi-mode resonant slotline antennas are expected to get an application in other novel multi-mode resonant antenna developments with controllable radiation dispersions.

3.3 OFFSET-FED MULTI-MODE RESONANT SLOTLINE ANTENNAS

3.3.1 Multi-Mode Resonant Slot Antenna with Frequency-Spatial Steerable Gain Notched Characteristic

In this section, the offset-fed technique can be employed to partially excite the even-order mode in addition to its odd-order counterparts. As is illustrated in Figure 3.20, for a horizontal slotline antenna, the resonant even-order modes contribute a radiation null in the bore-sight, $+y$-direction

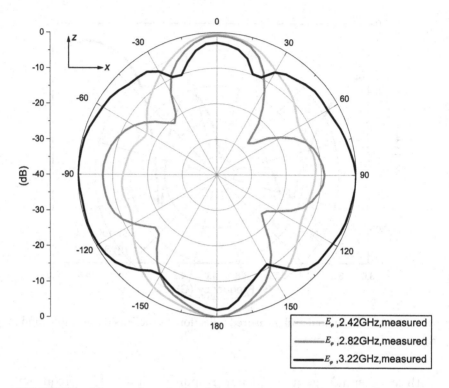

FIGURE 3.18   Measured xz-plane radiation patterns at 3.42, 3.82, and 4.22 GHz.

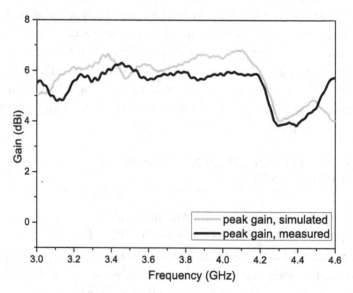

FIGURE 3.19   Simulated and measured peak gain of the triple-mode resonant slotline antenna.

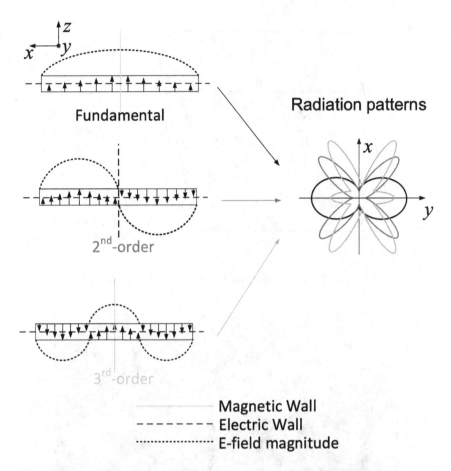

Fundamental

Radiation patterns

2nd-order

3rd-order

———— Magnetic Wall
– – – – – – – Electric Wall
··············· E-field magnitude

FIGURE 3.20 Electric field distributions, natural boundary conditions, and characteristic radiation patterns of the first three resonant modes in a slot-line radiator.

direction Xu et al. (2015). Thus, the radiation far-field superposition by odd- and even-order modes would possibly exhibit a gain notch characteristic. As revealed in Wang S. G. et al. (2017), frequency-spatial steerable gain notch functionality can be realized, while a continuous, wideband impedance frequency response is maintained.

The investigated, triple-mode resonant slotline antenna is designed on a square substrate with relative permittivity $\varepsilon_r = 2.65$, thickness $h = 1.0$ mm, with most parameters identical to that of Ant. 3 in Wang et al. (2017), with different $L_4 = 5$ mm and $L_3 = 12$ mm. The configuration and the corresponding fabricated prototype are shown in Figure 3.21a and b.

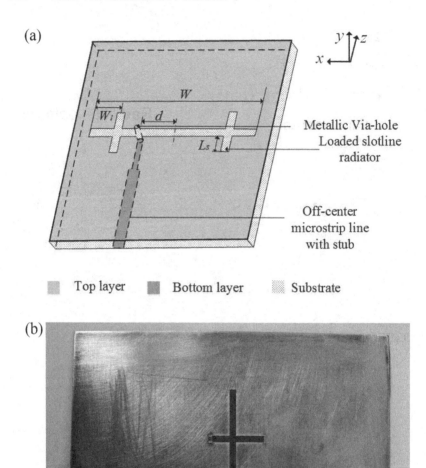

(a)

y ↑ z
x ←

W

W_1 d

L_s

Metallic Via-hole
Loaded slotline
radiator

Off-center
microstrip line
with stub

Top layer   Bottom layer   Substrate

(b)

FIGURE 3.21 Triple-mode resonant, frequency-spatial steerable gain notched slotline antenna: (a) configuration and (b) photograph of a fabricated prototype.

As has been revealed in Wang S. G. et al. (2017), the antenna can exhibit a wide, triple-mode resonant impedance bandwidth of over 40%. Figure 3.22 compares the measured peak gain and the bore-sight gain in +y-direction. As can be seen, when the even-order resonant mode has been excited, the peak gain can still maintain a relatively flat frequency response, which is quite similar to the all odd-order mode case in Guo et al. (2017). In the bore-sight direction, gain notch characteristic can be observed, which owns to the radiation of partially excited, full-wavelength mode, as predicted in Figure 3.20.

To further reveal the frequency-spatial dispersive behavior of the triple-mode resonant antenna, azimuth radiation patterns at different frequencies have been measured and compared in Figure 3.23. As can be seen, the excitation of even-order, a full-wavelength mode may lead to null frequency scanning phenomenon: The radiation null can be steered within an azimuth narrow angle range from $\varphi=65°$ to 75°. Similar to the case in Guo et al. (2017), the revealed operation principle can be employed to precisely adjust the radiation dispersion of multi-mode resonant antennas and to yield more novel designs with distinctive functionality.

The representative multi-mode resonant slotline antenna design in (Lu and Zhu 2015a) has been compared to a series of typical wideband slot

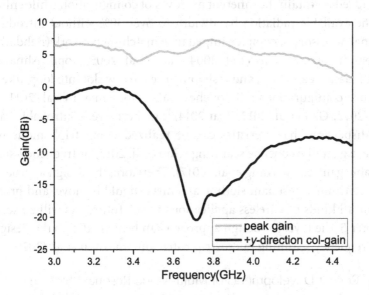

FIGURE 3.22 Peak and bore-sight (+y-direction), co-polarized gains in of the triple-mode resonant, frequency-spatial steerable gain notched slotline antennas.

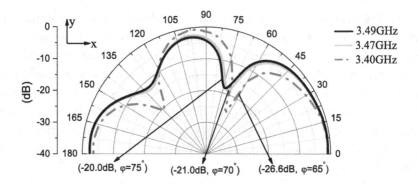

FIGURE 3.23   Measured H-plane (*xy*-plane) radiation pattern of the triple-mode resonant, frequency-spatial steerable gain notched slotline antenna at different frequencies.

antenna designs (Jang 2000, Sze and Wong 2001, Zhu et al. 2003, Behdad and Sarabandi 2004, Sharma et al. 2004, Latif et al. 2005, Gopikrishna et al. 2009, Huang et al. 2012, Kimouche et al. 2010, Samadi et al. 2011, Bod et al. 2012, Gao et al. 2013, Tsai 2014). By comprehensively comparing the results in (Lu and Zhu 2015b, Guo et al. 2017, Wang et al. 2017) and in this section, it can be concluded that the multi-mode resonant design approach can indeed maintain the inherent narrow slot configuration while enhancing the available radiation bandwidth to over 30% without introducing external accessories, complex impedance matching network (Behdad and Sarabandi 2004, Sharma et al. 2004, Latif et al. 2005, Gopikrishna et al. 2009, Huang et al. 2012), and reshaping the narrow slot into loop-like big-aperture configurations (Kimouche et al. 2010, Samadi et al. 2011, Bod et al. 2012, Gao et al. 2013, Tsai 2014). Furthermore, controllable radiation dispersion characteristics can be realized under triple-mode resonance, e.g., null frequency scanning (Guo et al. 2017) or frequency-spatial steerable gain notch Wang et al. (2017). Therefore, the design approach to 1-D multi-mode resonant slotline antennas should be novel and promising for all kinds of wireless applications in the future. As will be seen in Chapter 5, the revealed design approach can be extended to the design of null frequency scanning microstrip patch antennas Wu et al. (2020).

3.3.2 Recent Developments of Multi-Mode Resonant Slot Antennas

More recently, quad-mode resonant wideband characteristic has been implemented based on the multi-resonance concept Wang et al. (2017). The first four resonant modes can be excited by introducing an off-center feeding. As evidently presented in Wang H. et al. (2017), the radiation behaviors of the quad-mode resonant slot antenna well validate the property of each eigenmode within a single slotline radiator Wang et al. (2017), which is complementary to the one presented in the Appendix of Lu et al. (2018). As is expected, the presented quad-mode resonant design approach can be potentially employed in the compact mobile terminal antenna designs Wang et al. (2017).

The principle of the multi-mode resonant slotline antennas has also been further extended to design hand-held mobile terminal antennas (Pazin and Leviatan 2016, Ban et al. 2016, Wen and Hao 2016), mobile base station antennas Cai et al. (2018), Internet-of-Things sensor antennas (Seo et al. 2018), and broadband slotline switches (Ponchak 2016): The effect of the slot's position on impedance bandwidth of an edge-etched, multi-mode resonant slot antenna has been studied (Pazin and Leviatan 2016). The multi-mode resonant characteristic of a metallic frame tightly coupled and excited by a monopole has been studied and employed to realize a multi-band mobile phone antenna: Two shorting pins are properly incorporated into the segmented metallic frame, to adjust the resonant current's path and generate extra resonances Ban et al. (2016) for bandwidth enhancement. Modified dual-mode resonant slot antennas with diagonally loaded stubs have been used to design a two-port Multiple-Input-Multiple-Output (MIMO) antenna in mobile terminals. When the slotline radiator or the dipole is folded or meandered in arbitrary, it is found that the resultant antennas can still exhibit wideband, dual-mode resonant characteristics, and they can satisfy different applications in mobile communications and wireless sensors (Cai et al. 2018, Seo et al. 2018). This implies that the principle of multi-mode resonant slotline antennas should be valid to other complex antenna designs. More relevant examples for all kinds of applications will be raised and discussed in detail in Chapter 6.

3.4 BRIEF HISTORY AND RECENT DEVELOPMENTS OF LOOP ANTENNAS

As another type of widely recognized magnetic dipole, loop antennas have been intensively studied and applied in all kinds of radio systems since the late 1930s (Berndt and Gothe 1938). As is depicted in classical references (Kraus 1988), a horizontal small loop with uniform electric current distribution can be equivalently recognized as a z-oriented magnetic dipole. As shown in Figure 3.25a, a series of feeding and loading

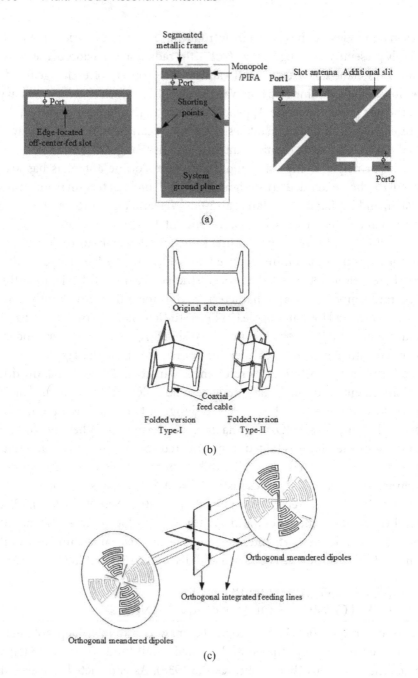

FIGURE 3.24   Recent applications of the principle of multi-mode resonant slot-line antennas: (a) mobile terminals (Pazin and Leviatan 2016, Ban et al. 2016, Wen and Hao 2016), (b) omnidirectional base station antennas (Cai et al. 2018), and (c) Internet-of-Things sensor antennas (Seo et al. 2018).

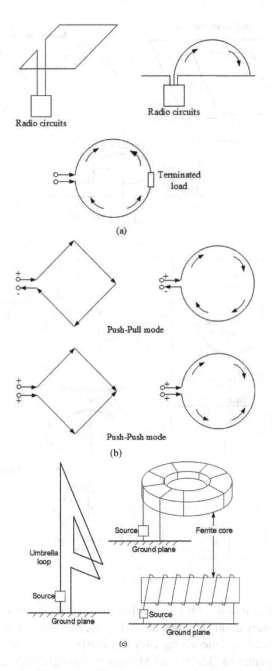

FIGURE 3.25 A brief history of loop antennas: (a) Short-wave loop antennas (Berndt and Gothe 1938, Alford and Kandoian 1940, Beverage 1941), (b) two distinctive operational modes of circular loop antennas (Schelkunoff and Friis 1952), (c) multi-turn, electrically small loop antennas (Fenwick 1965),

(Continued)

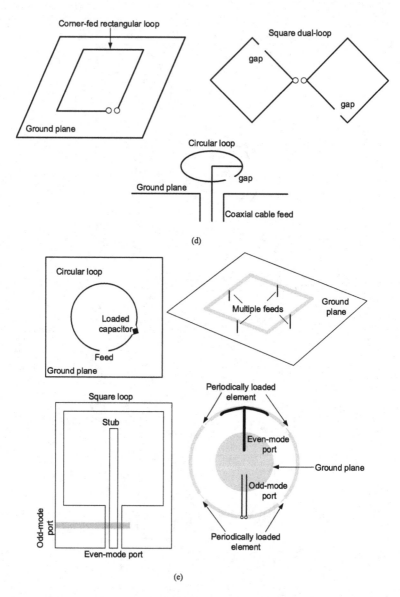

FIGURE 3.25 (*Continued*) (d) rectangular, rhombic and circular loop antennas for circular polarization (Murakarni et al. 1996, Morishita et al. 1998, Nakano 1998), (e) loaded loop antenna for circular polarization Li et al. (2005), pattern steerable loop antenna (Mehta and Mirshekar-Syahkal 2007), dual-mode loop antenna for diversity Li et al. (2011), and meta-inspired loop antenna Zhang et al. (2015).

techniques to classical short wave loop antenna has been developed, and loop antennas with different performances have been realized (Alford and Kandoian 1940, Beverage 1941). Since then, theoretical analyses on loop antennas have been intensively studied for many years (Sherman 1944, Schelkunoff and Friis 1952, Carter 1952, Lindsay 1960, Wu 1962, Prasad and Das 1970, Adekola 1983, Werner 2000, Balanis 2005). The mode theory of loop antenna was pioneered in (Schelkunoff and Friis 1952): The operational modes of the circular/square loop antennas can be depicted by "Push-Push" and "Push-Pull" modes, respectively, with the former mode dominated by cosine functions, and the latter dominated by sine ones, as shown in Figure 3.25b.

More mathematically elaborated models have been developed to calculate both the near and far fields of a thin loop antenna (Carter 1952, Lindsay 1960, Wu 1962, Prasad and Das 1970, Adekola 1983, Werner 2000). In many classical textbooks (Kraus 1988, Balanis 2005), an ideal loop antenna with pure standing-wave current distribution, horizontally placed in the xy-plane only generates E_φ component in the far-field, and the E-field along the loop's axis is always equal to zero regardless of its varied size (Kraus 1988, Balanis 2005). This conventional case can be accordingly called "Push-Pull mode" (Schelkunoff and Friis 1952). When the loop antenna exhibits a non-uniform, partial traveling-wave current distribution, the situation is quite different: Typically, when the perimeter of a loop is approximately approaching to one wavelength, its far-field would exhibit E_θ and E_φ components with radiation maxima occurred along the axis of the loop (Carter 1952, Lindsay 1960, Wu 1962, Prasad and Das 1970, Adekola 1983, Werner 2000). In this case, radiation behavior is alternatively dominated by the "Push-Push mode" (Schelkunoff and Friis 1952, Carter 1952, Adekola 1983) instead. The behavior and property of resonant loop antennas will be discussed in detail in the next section by studying the eigenmodes of a circular loop antenna.

Later on, loop antennas have been widely used to serve as electrically small antennas for reception (Kraus 1988, Fenwick 1965) in high-frequency radios. Multi-turn, wound loop antennas with different configurations have been developed, in which some of them are loaded by ferrite rods or cores (Fenwick 1965, Kraus 1988), and can be found in amplitude modulation radios, as shown in Figure 3.25c. Since the 1990s, loop antennas have been extensively employed to design all kinds of circularly polarized (CP) antennas (Murakarni et al. 1996, Morishita et al. 1998,

Nakano 1998), as shown in Figure 3.25d: A corner-fed rectangular loop backed by a metallic reflector can yield simple CP antenna with excellent performance Murakarni et al. (1996). If a pair of square loops is incorporated and properly loaded by open-circuited elements, wideband, unidirectional CP characteristics can be attained Morishita et al. (1998). A loop-monopole combined configuration mounted above a ground plane can lead to broadside (Nakano 1998) or conical Nakano et al. (2000) CP beams. By loading a lumped capacitor, wideband CP characteristic can be realized Li et al. (2005), as shown in Figure 3.25e. Multiple ports can be employed to symmetrically excite a square loop antenna backed by a ground plane. In this way, different operational modes can be excited and used to realize a reconfigurable, steerable pattern in the azimuth plane (Mehta and Mirshekar-Syahkal 2007) (Figure 3.25e). Since the 2010s, loop antennas have been drawn more and more attention owing to their potential for compact designs and polarization diversity: A diversity antenna using orthogonal polarizations can be implemented by a simple, rectangular loop-slot hybrid configuration Li et al. (2011), as illustrated in Figure 3.25e. With the development of periodic structures and metamaterials, meta-inspired loop antenna under zeroth-order resonance shown in Figure 3.25e has been advanced for size miniaturization and compact designs Zhang et al. (2015).

Based on the generalized odd-even mode theory, and the 1-D multi-mode resonant dipole theory, a multi-mode resonant loop antenna will be modeled and discussed in the next section. Then, the multi-mode resonant concept will be employed to design even-mode resonant square loop antennas.

3.5 LOOP ANTENNAS UNDER MULTIPLE-MODE RESONANCE

An ideal, infinitesimal circular loop antenna in free space shown in Figure 3.28a is studied. Suppose the loop antenna is placed in xy-plane with its center axis coinciding with the z-axis. As previously discussed in Section 1.2.1, the loop antenna can be recognized as a dipole antenna with both ends short-circuited and bent to coincide at $\varphi = \pm\pi$ (Schelkunoff and Friis 1952). Thus the eigenmode current distributions with associated short-circuited boundary condition should obey the telegrapher equation and should have the general solution shown in Eq. (3.1)

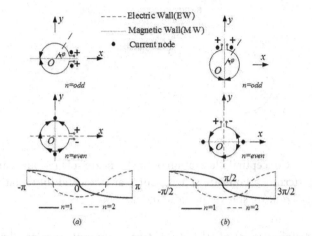

FIGURE 3.26 Resonant current eigenmodes in an ideal circular loop antenna: (a) symmetric about the x-axis ($\varphi = 0$) and (b) center rotated case, symmetric about the y-axis ($\varphi = 90°$).

$$\left.\begin{array}{l} \left(\dfrac{d^2}{d\phi^2}+k_\phi^2\right)I_\phi(\phi)=0 \\[3mm] \left.\dfrac{\partial I_\phi(\phi)}{\partial\phi}\right|_{\phi=\pi}=0 \\[3mm] \left.\dfrac{\partial I_\phi(\phi)}{\partial\phi}\right|_{\phi=-\pi}=0 \end{array}\right\} \Rightarrow I_\phi(\phi)= \begin{cases} J_0\sin\dfrac{n}{2}\phi, n=odd \\[3mm] J_0\cos\dfrac{n}{2}\phi, n=even \end{cases} \tag{3.1}$$

$$\left.\begin{array}{l} \left(\dfrac{d^2}{d\phi^2}+k_\phi^2\right)I_\phi(\phi)=0 \\[3mm] \left.\dfrac{\partial I_\phi(\phi)}{\partial\phi}\right|_{\phi=-\frac{\pi}{2}}=0 \\[3mm] \left.\dfrac{\partial I_\phi(\phi)}{\partial\phi}\right|_{\phi=\frac{3\pi}{2}}=0 \end{array}\right\} \Rightarrow I_\phi(\phi)= \begin{cases} J_0\cos\dfrac{n}{2}\left(\phi+\dfrac{\pi}{2}\right), n=odd \\[3mm] J_0\sin\dfrac{n}{2}\phi, n=even \end{cases} \tag{3.2}$$

$$I_\phi(\phi)=J_0\cos\phi\pm jJ_0\sin\phi=J_0 e^{\pm j\phi} \tag{3.3}$$

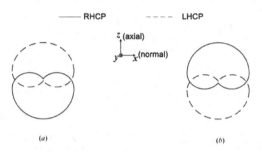

FIGURE 3.27 Conceptual illustration of the Poincaré sphere source antenna: (a) radiation pattern of a Poincaré sphere source antenna and (b) radiation pattern of an inverse Poincaré sphere source antenna.

Current distributions of the first two resonant modes depicted by Eq. (3.1) and the corresponding natural boundary conditions are shown in Figure 3.26a. Unlike the straight dipole in Chapters 1 and 2, a circular loop antenna is not only axially symmetric about $\varphi=0$ (i.e., x-axis), but also rotationally symmetric about the center of O. Therefore, if the loop is orthogonally rotated about the center, the sine- and cosine-dependency of eigenmodes should interchange, which implies that the resonant current eigenmodes of a 90°-rotated circular loop shown in Figure 3.26b should satisfy Eq. (3.2). As observed from Eq. (3.1) and Eq. (3.2), it is seen *the even-order resonant modes within an ideal, centro-symmetric, infinitesimal circular loop should always exist in pairs, and they should be orthogonal, degenerate mode pairs with an identical resonant frequency.*

Let's continue to study the case for $n=2$, i.e., the lowest order, even-mode resonance. Suppose the two degenerate, even-modes can be simultaneously exited with equal magnitude. When a 90° relative leading/lagging phase shift can be introduced by suitable perturbations (Wolff 1972), an exponential-dependent current distribution (mathematically equivalent to the traveling-wave one) shown in Eq. (3.3) will be attained. In this manner, a bidirectional, bisensing CP radiation characteristic Chen et al. (2017) will be resulted in, with elevational radiation patterns in xz- and zy-cut planes shown in Figure 3.27. The bidirectional, bisensing CP radiation patterns can also be approximately estimated by decomposing the loop current distribution into segmented current sheets Chen et al. (2017). It is found that the even-mode resonant circular loop cavity-backed by a metallic reflector should be an axial-mode, CP radiator, which is quite similar (but not equivalent) to the axial-mode, Kraus helical antenna with CP radiations (Kraus 1988). As is seen, the lowest order, even-mode resonant loop antenna radiates pure right-handed circularly polarized (RHCP) and right-handed elliptically polarized (RHEP) waves in the north pole and the north semisphere, respectively, while it radiates pure left-handed circularly polarized (LHCP) and left-handed elliptically polarized (LHEP) ones in the south pole and the south semisphere instead. In the equatorial plane, the antenna radiates linearly polarized (LP) waves. Indeed, by modifying the 90° leading/lagging phase shift, the polarization of the radiation pattern can be interchanged.

Such bidirectional, bisensing, and interchangeable radiation behavior shown in Figure 3.27 intuitively and vividly illustrates the concept of Poincaré sphere for partially polarized field representation (Deschamps 1951, Deschamps and Mast 1973). In this opinion, an ideal, bidirectional, even-mode resonant, dual CP loop antenna with LHCP/RHCP waves radiated in the north/south poles, LHEP/RHEP waves in the north/south semispheres, and LP waves in the equatorial plane could be reasonably named as the "*Poincaré sphere source antenna*", while its inverse counterpart (i.e., with RHCP/LHCP waves radiated in the north/south poles, RHEP/LHEP waves in the north/south semispheres, and LP waves in the equatorial plane) should be accordingly called as the "*Inverse Poincaré sphere source antenna*", as shown in Figure 3.27a and b (Chen et al. 2017).

The eigenmode theory for multi-mode resonant circular loop antenna is also valid for square loop antenna and other centro-symmetric loop

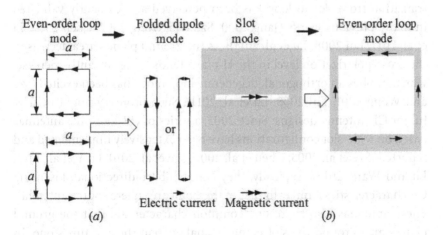

(a)   (b)

FIGURE 3.28 Transitions between slot, loop, and folded-dipole modes: (a) mode transition from loop to folded-dipole and (b) mode transition from slot to even-mode resonant loop.

antennas, irrespective of the shape of the loop. In those cases, the eigen-functions would be more complicated but still obey the "sine-cosine/cosine-sine alternating discipline" (Leontovich and Levin 1944) rediscovered in Chapter 1, while the basic radiation behavior is less dependent on the shape of the loop as discussed in (Schelkunoff and Friis 1952, Balanis 2005). Therefore, we can alternatively study and design a square loop antenna based on the multi-mode resonant theory deduced from a circular loop.

Furthermore, we can use the multi-mode resonant theory to illustrate the relationship and mode evolution between loop, dipole, and slot antennas in an intuitive manner. As shown in Figure 3.28a, when the width-to-length ratio of the square loop is adjusted, the excitation of a pair of even-modes can be accordingly tuned. When the square loop is modified into a slim rectangular one with an extremely high width-to-length ratio, one of the degenerate modes would be sufficiently suppressed, and the loop would accordingly evolve into a folded dipole (Harrison and King 1961).

Based on the previous discussions, let's continue to reveal the evolution between the slot antenna and loop antenna shown in Figure 3.28b. When the slot antenna's ground plane is drastically shrunk, and the width-to-length ratio of the slot is approaching 1:1, a pure "slot mode" would gradually transit to a "loop mode" under even-mode resonance: The electric current concentrated on the aperture's edge will dominate the radiation behavior, and the equivalent magnetic current mode within the slot will become less dominant accordingly. The phenomenon of mode transition from slot to loop has been perceived and evidently validated in many previous works (Jang 2000, Sze and Wong 2001, Chen 2003, Li et al. 2011, Lui 2008, Lu et al. 2011): As the ground plane's area decreases, the cross-polarization level in the H-plane tends to significantly increase, which implies an orthogonal, degenerate loop mode has been excited (Sze and Wong 2001, Lui 2008, Lu et al. 2011). Such property might be useful for CP antenna designs: Since 2003, a series of CP, loop-like antennas based on wide-slot configurations have been extensively implemented and reported (Sze et al. 2003, Chen et al. 2006, Sze et al. 2010, Lu Y et al. 2012, Lu and Wang 2013). Typically, they can exhibit bidirectional, bisensing CP characteristics similar to that of the Poincaré sphere source antennas. These antennas also share one common characteristic that the ground plane's area around the slot is much smaller than the aperture's one. In this opinion, the "slot mode" has partially been evolved into "loop mode" (under even-mode resonance), and the resultant antennas should operate under the complicated, hybrid resonances of slot mode, loop mode, and monopole mode (which is dominated by the tuning stub) Pan et al. (2014).

In the next section, the concept of Poincaré sphere source antennas will be experimentally validated by designing the simplest, even-mode resonant, CP square loop antennas. Several examples, including unbalanced and balanced cases, will be raised and discussed in detail.

3.6 POINCARÉ SPHERE SOURCE ANTENNAS: THE SIMPLEST, DUAL EVEN-MODE RESONANT CIRCULARLY POLARIZED LOOP ANTENNAS

3.6.1 Dual Circularly Polarized Loop Antenna

An even-mode resonant circularly polarized square loop antenna can be designed by simultaneously exciting a pair of even-mode with equal magnitude and 90° relative leading/lagging phase shift. As can be seen from Figure 3.26 and Eqs. (3.1)–(3.3), a diagonal feeder should be introduced to attain the equal-magnitude condition for two orthogonal polarizations. To satisfy the phase quadrature condition, the orthogonal currents' relative leading/lagging phase shift can be finely tuned by controlling the adjacent edges' width ratio (Wheeler 1950). Therefore, an original prototype design can be attained Zhang et al. (2016), as shown in Figure 3.29. The loop antenna is fed at the lower right corner by an open-circuited, fork-shaped (Jang 2000) microstrip stub. The non-uniform square loop is designed on a substrate with relative permittivity of $\varepsilon_r = 2.65$, tan $\delta = 0.001$, and thickness $h = 1.0$ mm, with a perimeter of about 1.08 guided-wavelengths at 2.45 GHz. As qualitatively described in (Wheeler 1950), the led/retarded phase δ_0 should be dependent on the loop's thickness. Thus the ratio of adjacent edges' widths defined as $d = S_1/S_2$ have been studied and illustrated in Figure 3.30.

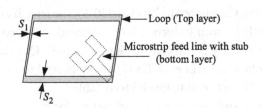

FIGURE 3.29 Conceptual design of corner-fed, CP loop antenna under dual, even-mode resonance.

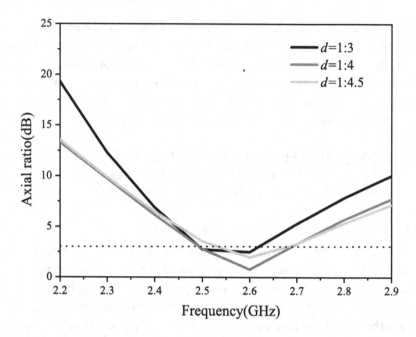

FIGURE 3.30 Simulated effect of d on the axial ratio (AR) in the axial direction ($\theta=0°$).

As reported in Zhang et al. (2016), a uniform ($d=1$) loop should exhibit linearly polarized in the axial-direction. Figure 3.30 shows the axial ratio (AR) in axial-direction varying with d: With d increases to 1:3, the AR in the axial-direction can be tuned down to 2.5 dB. When d is 1:4, a 3 dB AR bandwidth from 2.50 to 2.70 GHz with the lowest AR of 0.8 dB can be attained. With d increasing to 1:4.5, the CP bandwidth decreases again.

Based on the preliminary studies, a dual CP square loop antenna can be constructed on a dielectric substrate with relative permittivity of $\varepsilon_r=2.65$, tan $\delta=0.001$, and thickness $h=1.0$ mm. The inner perimeter is set to be about one guided-wavelength at 2.20 GHz, with key parameters tabulated in Xu et al. (2019). Compared to Xu et al. (2019), the length of lower corners has been extended 2 mm to form a larger ground plane, so that a coaxial cable can be vertically connected to the feed point. Quarter-wavelength, cylindrical choke sleeves Li et al. (2017) are employed to suppress the spurious radiation from the unbalanced feed cable.

The measured and simulated S-parameters for both ports are plotted in Figure 3.32. As can be seen, for both ports, the antenna can exhibit an impedance bandwidth for $|S_{11}|$ lower than −10 dB and port isolation

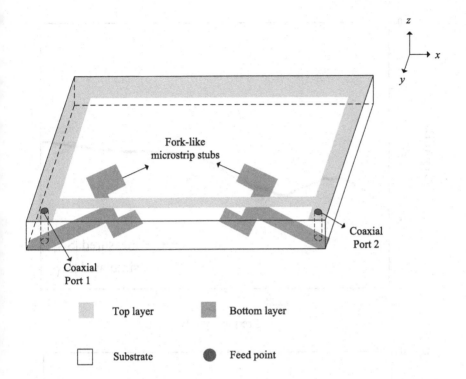

FIGURE 3.31   Configuration of the dual CP loop antenna under even-mode resonance.

for $|S_{21}|$ lower than −15 dB from 2.19 to 2.42 GHz, which is about 10% in fraction. The AR frequency responses in axial, +z-direction for both ports are measured and shown in Figure 3.33. As can be seen, the 3 dB AR bandwidth for both ports can be determined as 2.11–2.41 GHz, which is approximately identical to the impedance one.

The beam symmetry has been validated at 2.20 GHz in Xu et al. (2019). Therefore, only one principal-cut plane is presented in this section. As shown in Figure 3.34, the radiation patterns at both ports in the xz-plane at 2.18 GHz also exhibit good bidirectional CP characteristics. When the antenna is excited at Port 1, a Poincaré sphere source antenna can be yielded, and an inverse one can be attained when Port 2 is excited.

Figure 3.35 shows the simulated and measured gains in the axial-direction at both ports. The average gains for Port 1 and Port 2 are 2.7 and 2.3 dBic, respectively. To enhance the gain in the axial direction, the Poincaré sphere source antenna should be mounted above a metallic reflector to yield a unidirectional version shown in Figure 3.36.

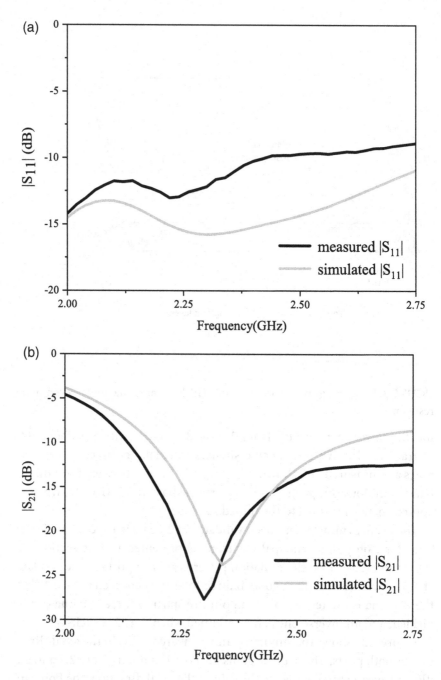

FIGURE 3.32 Measured and simulated S-parameters of the bidirectional dual CP loop antenna: (a) $|S_{11}|$ and (b) $|S_{21}|$.

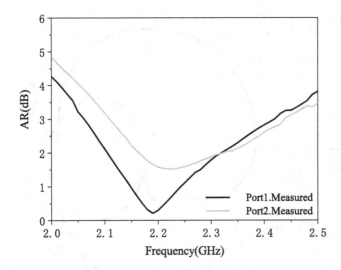

FIGURE 3.33   Measured AR frequency responses of the bidirectional antenna: (a) Port 1 and (b) Port 2.

As is shown in Figure 3.37, the reflection coefficients for both ports tend to exhibit wider −10 dB bandwidths, as well as the port isolations behave, than the bidirectional, reflector free case. The impedance bandwidth for both ports can be determined as 2.19–2.57 GHz, i.e., 15.2% in fraction. Figure 3.38 shows the measured AR frequency response of the unidirectional version, where the upper limit of 3 dB AR bandwidth for Port 1 is limited to 2.38 GHz. By considering the impedance bandwidth and upper 3 dB AR frequency, the available CP radiation bandwidth of the unidirectional antenna can be concluded as from 2.19 to 2.38 GHz, i.e., 8.5% in fraction.

Figure 3.39 shows the measured radiation patterns in xz-plane at Port 1 at 2.21 GHz and Port 2 at 2.18 GHz. The antenna exhibits unidirectional CP radiation patterns with a front-to-back ratio (FBR) of over 27 dB at both ports. Figure 3.40 shows the gains in the axial direction of the unidirectional antenna. At both ports, the average gains in axial-direction are up to 7.9 and 7.7 dBic, respectively. Compared to the bidirectional case, gain enhancement for both ports is over 5 dB.

In this section, the design approach to a novel dual CP square loop antenna is presented. As validated, the conceptual Poincaré sphere source antenna and its inverse counterpart can be co-located excited, implemented, and used as a polarization diversity antenna Xu et al. (2019). Both implemented bidirectional and unidirectional prototypes have been

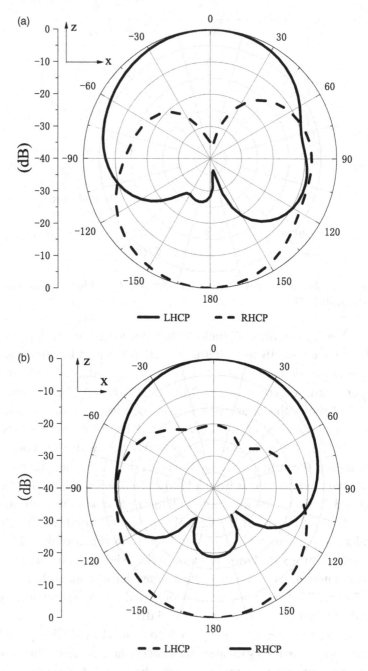

FIGURE 3.34   Measured radiation patterns of the proposed antenna: (a) *xz*-plane at Port 1, 2.18 GHz and (b) *xz*-plane at Port 2, 2.18 GHz.

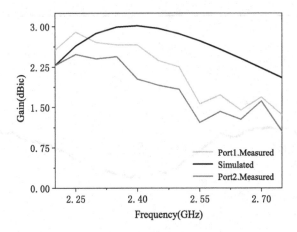

FIGURE 3.35 Simulated and measured gain frequency responses in the axial direction ($\theta = 0°$).

designed and excited by the simplest choke sleeve coaxial cables. The operation principle of Poincaré sphere source antenna can be extended to get applications in the design of polarization diversity antenna systems or polarization agile antenna systems.

3.6.2 Balanced Circularly Polarized Loop Antenna

The Poincaré sphere source antennas presented in the previous section are all unbalanced, single-ended-fed versions. With reference to the natural boundary condition revealed in Chen et al. (2017) and Figure 3.26, balanced Poincaré sphere source antennas will be presented and discussed in this section.

FIGURE 3.36 Photograph of a fabricated prototype of unidirectional, Poincaré sphere source antenna mounted above a metallic reflector.

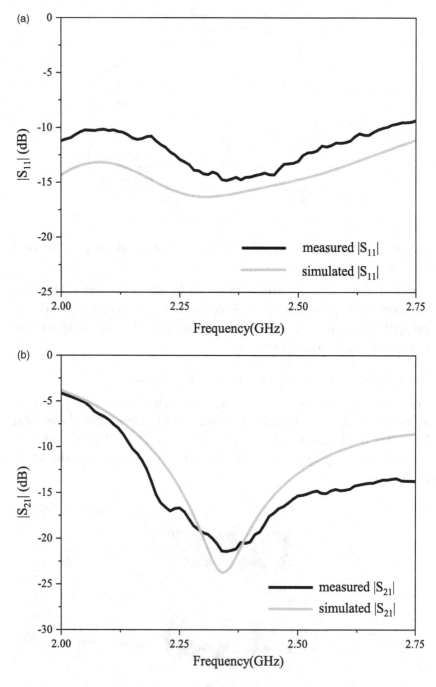

FIGURE 3.37 Measured and simulated S-parameters of the unidirectional, dual CP antenna: (a) $|S_{11}|$ and $|S_{21}|$.

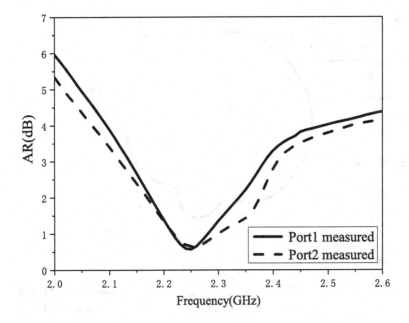

FIGURE 3.38   Measured AR frequency responses of the unidirectional version in the axial direction ($\theta=0°$).

The configuration of a balanced Poincaré sphere source antenna is illustrated in Figure 3.41a. As can be seen, the loop radiator is diagonally excited by a dipole launcher, with its center connected to a balun. All design parameters are identical to the microstrip balun fed case, which has been intensively illustrated in Yuan et al. (2018). Herein, the loop antenna under investigation will be excited by a rigid coaxial cable with a Pawsey-stub (Cork and Pawsey 1939) instead, for reliable mechanical support in practice. Figure 3.41b shows the photograph of a fabricated prototype with a Pawsey-stub that serves as a balun transformer from twin-wire to coaxial cable.

Figure 3.42 compares the simulated reflection coefficient without balun to the measured, Pawsey-stub one. As validated in Lu, Zhang et al. (2014), the antenna's inherent resonant characteristic is seldom affected by the external balun: The discrepancy between simulated and measured center frequencies is less than 1%, with the impedance bandwidths for reflection coefficient lower than −10 dB reasonably agree. Compared to the microstrip balun case in Yuan et al. (2018), the measured impedance bandwidth is narrower but is still as wide as 13% owing to the dipole launcher with fork-shaped stubs (Sze and Wong 2001, Zhang et al. 2016).

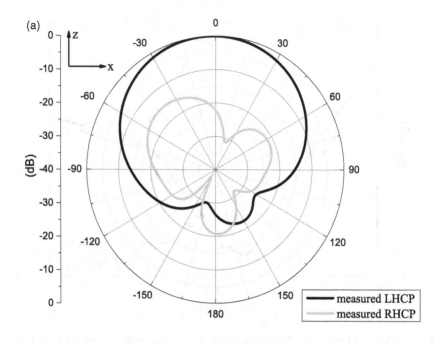

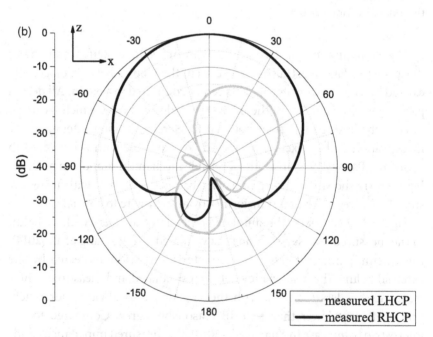

FIGURE 3.39 Measured radiation patterns of the unidirectional antenna: (a) *xz*-plane at Port 1, 2.21 GHz and (b) *xz*-plane at Port 2, 2.18 GHz.

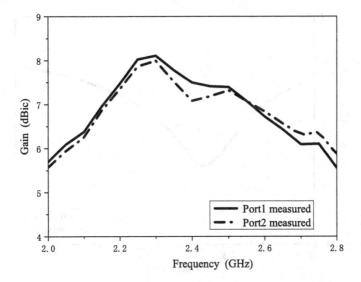

FIGURE 3.40   The measured gains in the axial direction ($\theta=0°$) of the unidirectional antenna.

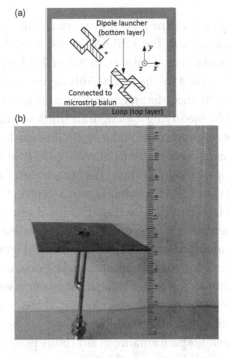

FIGURE 3.41   Balanced even-mode resonant, CP loop antenna: (a) configuration and (b) photograph of a fabricated prototype with a Pawsey-stub.

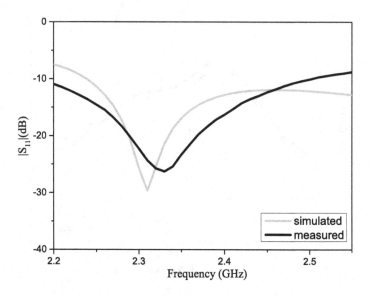

FIGURE 3.42 Measured and simulated reflection coefficients of the loop antenna with a Pawsey-stub balun.

As shown in Figure 3.43, the loop antenna can exhibit a good CP performance in the axial direction. The 3 dB AR bandwidth is about 6.7% varying from 2.32 to 2.48 GHz, with a minimum AR of 2 dB. The degradation of CP performance is caused by the balun. Although a microstrip balun may offer better CP performance, the antenna's CP performance is expected to be improved by carefully modifying the length and thickness of the Pawsey-stub.

Figure 3.44 plots the measured and simulated normalized radiation pattern at 2.41 GHz. As can be seen, the balanced version Poincaré sphere source antenna exhibits similar bidirectional CP characteristics as its unbalanced counterpart in Xu et al. (2019) behaves. Figure 3.45 gives the axial-direction gain of the antenna, with in-band average measured gain of 3.5 dBic.

The balanced bidirectional antenna is further installed above a metallic reflector to yield a unidirectional version, as shown in Figure 3.46. As can be observed from Figure 3.47, in the reflector-backed case, a Pawsey-stub may cause frequency shift to the center resonant frequency, with a 5% discrepancy to the simulated, microstrip balun fed case. Therefore, the length of the Pawset-stub's (possibly, including its thickness, too) can be carefully adjusted to optimize the antenna's performance in any future practical developments.

Figure 3.48 illustrates the measured AR frequency response and compares it to the simulated, microstrip balun fed case. It is found the AR frequency

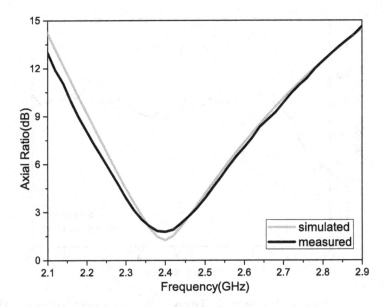

FIGURE 3.43 Measured and simulated axial ratios of the bidirectional antenna in the axial direction ($\theta = 0°$).

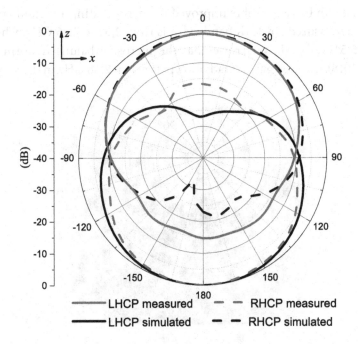

FIGURE 3.44 Simulated and measured normalized radiation patterns of the bidirectional, balanced loop antenna, xz-plane, 2.41 GHz.

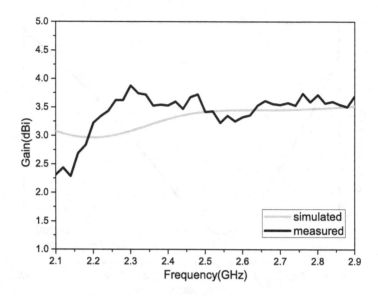

FIGURE 3.45   Simulated and measured gains in the axial-direction ($\theta = 0°$) of the bidirectional, balanced loop antenna.

response can be remarkably improved by incorporating a metallic reflector. The measured 3 dB AR bandwidth is from 2.32 to 2.45 GHz, which is about 5.5% and slightly narrower than the microstrip balun fed case in Yuan et al. (2018). The normalized radiation patterns at 2.38 GHz in the xz-plane

FIGURE 3.46   Photograph of a fabricated prototype of unidirectional, balanced CP loop antenna.

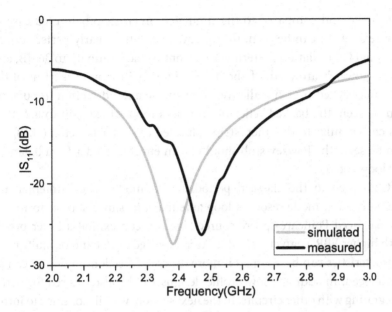

FIGURE 3.47 Measured and simulated reflection coefficients of the unidirectional, balanced CP loop antenna.

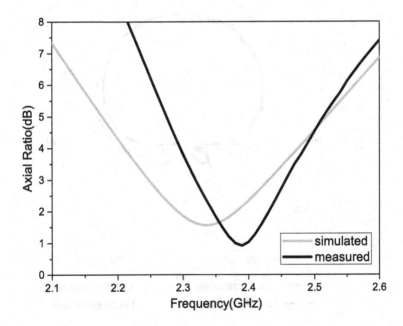

FIGURE 3.48 Simulated and measured axial ratios of the unidirectional, balanced CP loop antenna in the axial direction ($\theta = 0°$).

are shown and compared to the simulated, microstrip balun fed case in Figure 3.49. As can be seen, the antenna exhibits a nearly perfect unidirectional CP radiation pattern with a front-to-back ratio up to 36 dB, and a half-power beamwidth of about 70°. Figure 3.50 shows the gains of the antenna backed by a metallic reflector in the axial direction ($\theta=0°$). As can be seen, the backed reflector enhances the average gain to 8.2 dBic, which is similar to the microstrip balun fed case in Yuan et al (2018). As can be seen, the Pawsey-stub should be an effective feeder for a balanced CP loop antenna.

Compared to the classical periodic, axial-mode helical antenna, the presented even-mode resonant loop antenna at its simplest basic form, i.e., the balanced Poincaré sphere source antenna, can exhibit a lower profile with higher FBR Yuan et al. (2018). As is expected, wideband or multi-band characteristics can be attained if more degree of freedom in design can be introduced. In addition, low-profile designs will be also desired for better integrating with other circuits. In the next section, we will continue to introduce the relevant design techniques of the Poincaré sphere source antenna.

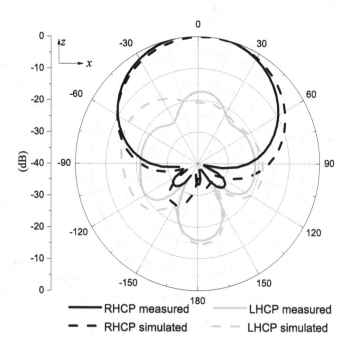

FIGURE 3.49 Simulated and measured normalized xz-plane radiation patterns of the unidirectional, balanced CP loop antenna at 2.38 GHz.

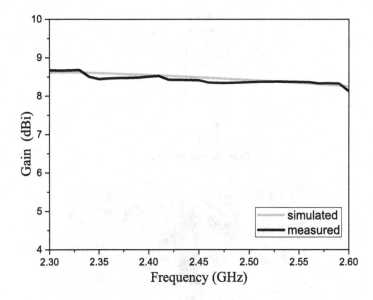

FIGURE 3.50   Simulated and measured gains in the axial direction ($\theta = 0°$) of the balanced CP loop antenna.

3.6.3 Wideband and Multi-Band Even-Mode Resonant Circularly Polarized Loop Antennas

In the previous sections, the simplest even-mode resonant CP loop antenna named Poincaré sphere source antenna, including both balanced and unbalanced cases have been studied. As is seen, their AR bandwidth is single-mode resonant and limited to less than 10%. Therefore, wideband designs with multi-mode resonant AR frequency response have been studied (Liu et al. 2018, Liu 2019).

Figure 3.51a illustrates the configuration of a CP, wideband dual-loop antenna. As can be seen, the two coaxially embedded, rectangular loops are diagonally excited by a dipole launcher on the same layer. The inner and outer perimeters of the loop are 0.85 and 1.18 guided wavelengths at 2.45 GHz, respectively. Detailed design parameters of the antenna have been given in Liu et al. (2018). Figure 3. 51b shows the photograph of a fabricated prototype.

The simulated and measured reflection coefficients are shown in Figure 3.52a. The coaxially embedded, dual-loop antenna exhibits a wideband, multi-mode resonant characteristic from 2.36 to 4.0 GHz, with fractional impedance bandwidth for $|S_{11}|$ lower than −10 dB up to 51.6%. As illustrated in Figure 3.52b, the measured AR bandwidth for AR smaller

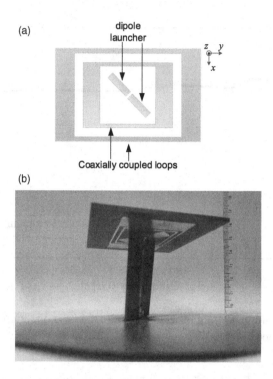

FIGURE 3.51 Coaxially embedded, CP dual-loop antenna: (a) configuration and (b) photograph of a fabricated prototype.

than 3 dB ranges from 2.43 to 2.85 GHz, i.e., about 16% in fraction. The simulated and measured xz-plane radiation patterns at 2.45 GHz are presented in Figure 3.53. As can be seen, the experimental results agree well with the numerical ones. The FBR of the antenna is as high as 30 dB, which implies that the wideband antenna should exhibit good LHCP CP purity. The simulated and measured gains in the axial direction are plotted in Figure 3.54. The in-band average gain is about 5 dBi, which is slightly inferior to that of the narrowband design in Yuan et al. (2018). As observed from the impedance matching, AR, radiation pattern and gain, the available bandwidth of the CP dual-loop antenna should be determined as 16%, which is doubled compared to the single-loop case Yuan et al. (2018). This convinces that the coaxially embedded dual-loop concept should be effective for AR bandwidth enhancement of even-mode resonant loop antennas.

Besides the wideband design, multi-band designs can be obtained by simultaneously exciting the first ($n=2$) and the third ($n=6$) pairs of even-mode (Wang et al. 2017, Zhang et al. 2019, Zhang 2020). As conceptually

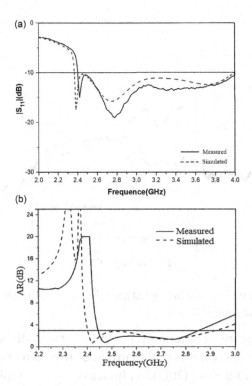

FIGURE 3.52 Simulated and measured impedance and AR bandwidth of the wideband CP dual-loop antenna: (a) reflection coefficients and (b) AR.

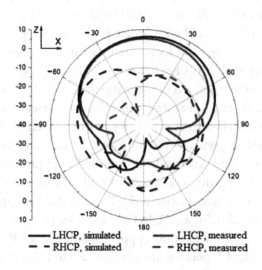

FIGURE 3.53 Simulated and measured xz-plane radiation patterns at 2.45 GHz.

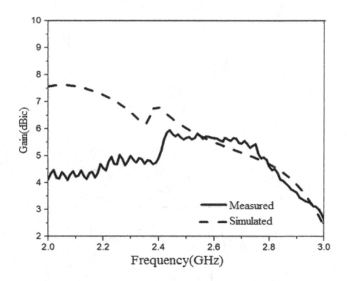

FIGURE 3.54 Simulated and measured gains in the axial direction ($\theta = 0°$) of the wideband CP dual-loop antenna.

illustrated in Figure 3.55, four stubs can be rotationally and symmetrically introduced at the nodes of the current distribution of the third-order even mode Lu, Zhu et al. (2017). As indicated by the empirical formulas in (Wang et al. 2017, Zhang et al. 2019), the dual-band ratio can be tuned from 1.4:1 to 1.7:1 by modifying the length, shape, and position of the four stubs (Wang et al. 2017, Zhang et al. 2019, Zhang 2020). A pair of orthogonal dipole launchers with fork-like stubs is employed to excite the loop radiator. A balanced dual-band design is shown in Figure 3.56, Design parameters of the dual-band loop antennas can be a reference to (Wang et al. 2017, Zhang et al. 2019). For better dual-band CP operation, additional digits have been incorporated into one dipole launcher (Zhang 2020).

The simulated and measured reflection coefficients are shown in Figure 3.57a. As can be seen, the stub-loaded loop antenna exhibits a dual-band characteristic from 1.4–1.86 to 2.59–3.09 GHz, for $|S_{11}|$ lower than −10 dB. The measured AR bandwidth plotted in Figure 3.57b is inferior to the simulated one, which ranges from 1.49–1.51 to 2.45–2.48 GHz, i.e., 1.3% and 1.2% in fraction, respectively. The simulated and measured radiation patterns at 1.5, 1.54, and 2.48 GHz are presented in Figure 3.58. As can be seen, the antenna exhibits opposite handedness CP characteristics at both bands. The CP performance is sensitive to the additional digit of D_7. The simulated and measured gains in the axial direction ($\theta = 0°$) are

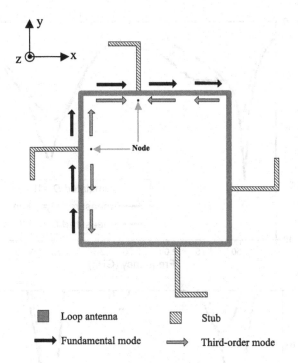

FIGURE 3.55 Conceptual stub-loaded, dual-band CP loop antenna design.

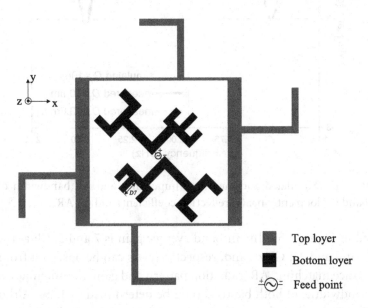

FIGURE 3.56 Dual-band CP loop antenna loaded with stub and excited by dual dipole launchers.

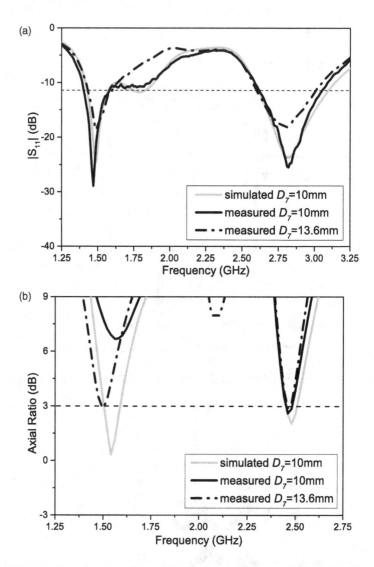

FIGURE 3.57 Simulated and measured impedance and AR bandwidth of the dual-band CP loop antenna: (a) reflection coefficients and (b) AR.

plotted in Figure 3.59. The in-band average gain is 7 and 5 dBi at the 1.5 GHz-band and 2.4 GHz-band, respectively. As can be observed from the impedance matching, AR, radiation pattern and gain, the antenna's available bandwidths of both bands should be determined as 1.3%. Although the experimental performance is inferior to the numerical one, it can be further optimized in the future.

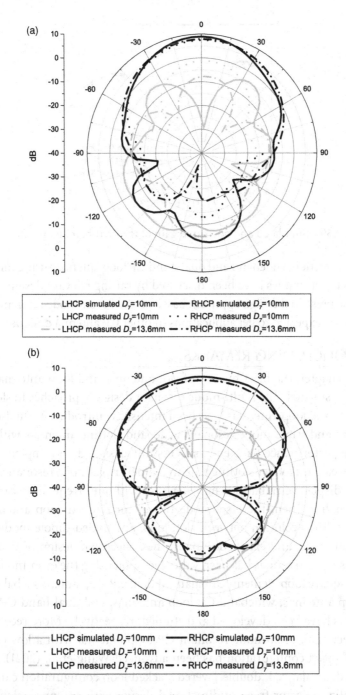

FIGURE 3.58 Simulated and measured *xz*-plane radiation patterns: (a) $D_7 = 10$ mm: 1.54 GHz; $D_7 = 13.6$ mm: 1.5 GHz and (b) 2.48 GHz.

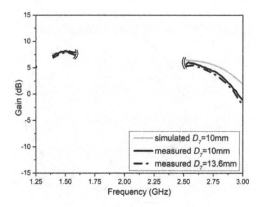

FIGURE 3.59 Simulated and measured gain of the dual-band loop antenna.

In this section, wideband and dual-band CP loop antennas using multiple resonant even-modes have been discussed by raising relevant design examples. The presented design approach based on dual even-mode resonant loop elements is expected to get applications in other CP antenna designs.

3.7 CONCLUDING REMARKS

In this chapter, the brief history of slot antennas and loop antennas has been investigated. The multi-mode resonant design approaches to slotline antennas and loop antennas have been intensively introduced. Dual-mode resonant and triple-mode resonant wideband slotline antennas with distinctive performance have been successfully designed and implemented. As is revealed, a longer slotline antenna under triple-mode resonance may exhibit dispersive radiation characteristics with null frequency scanning functionality. Then, the eigenmode theory for circular loop antenna is deduced. The evolution between slot, loop, and folded-dipole modes has been illustrated and clarified. Finally, the conceptual design of Poincaré sphere source antennas has been validated upon using the even-mode resonant, square loop element: Unbalanced dual CP loop antennas, balanced CP loop antennas, wideband CP loop antennas, and dual-band CP loop antennas have been developed and studied, respectively. More recently, a unidirectional, even-mode resonant CP loop antenna with a low profile and balanced configuration has been advanced Wang et al. (2021). As is reported, a balanced, double-layered stacked loop configuration can lead to a bulky reflector-free, unidirectional design with antenna height less than 0.1-wavelength. It is expected more novel designs could be developed from the simplest, multi-mode resonant slot/loop antennas.

CHAPTER 4

Multi-Mode Resonant Complementary Dipole Antennas

4.1 BRIEF HISTORY AND RECENT DEVELOPMENTS OF COMPLEMENTARY DIPOLE ANTENNAS

In the previous chapters, we have discussed basic electric and magnetic elementary dipoles in the opinion of multi-mode resonance. In this chapter, we will continue to discuss the complementary dipole antennas, which can be recognized as the simplest combined elementary antenna made up of electric and magnetic dipoles. Complementary dipole antennas containing an electric dipole and a magnetic dipole are also known as the "Huygens source antennas", with their history dated back to the late 1940s (Wheeler 1947, Kandoian 1951, Clavin 1954, King and Owyang 1960, Mayes et al. 1971). Originally, complementary dipole antenna should be composed of a horizontally polarized loop and a vertically polarized dipole element (Figure 4.1a), with both elements co-located excited with equal-magnitude and 90° phase shift. This yields circularly polarized (CP), electrically small helical antenna (Wheeler 1947), or complementary dipole antenna for improvement of frequency modulation broadcasting coverage (Kandoian 1951). Later, a waveguide-based complementary dipole antenna was developed by incorporating a dipole to a TE_{11} magnetic source in a circular waveguide (Clavin 1954) (Figure 4.1b), to achieve equal E- and

DOI: 10.1201/9781003291633-4

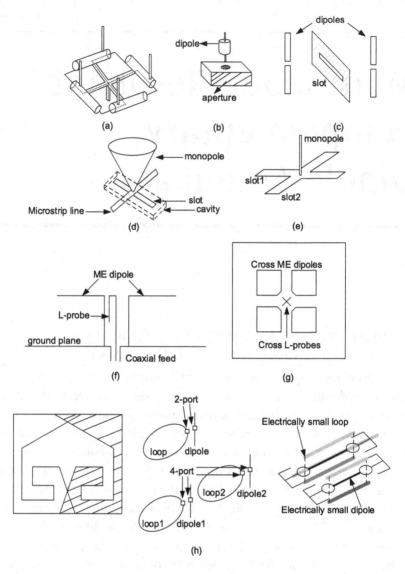

FIGURE 4.1   Classical and advanced complementary dipole antennas: (a) circularly polarized loop-dipole antenna (Kandoian 1951), (b) pencil-beam feed (Clavin 1954), (c) coupled slot-dipole antenna (King and Owyang 1960), (d) monopole-slot combined antenna (Mayes et al. 1971), (e) cross-slot-monopole antenna (Itoh and Cheng 1972), (f) magnetoelectric dipole antenna (Luk and Wong 2006), (g) dual-polarized magnetoelectric dipole antenna (Luk and Wu 2012), and (h) electrically small Huygens source antennas (Lu et al. 2010, Best 2010, Tang et al. 2019).

H-plane and to serve as a pencil-beam feed for parabolic reflector antennas. Multiple passive dipoles or monopole(s) can be incorporated into a slot element (King and Owyang 1960, Mayes et al. 1971, Itoh and Cheng 1972), and a series of complementary dipole antenna with wideband characteristics would be yielded and implemented (Figure 4.1c–e).

As is seen, most of the early designs exhibit narrowband performance (Kandoian 1951, Clavin 1954, King and Owyang 1960) or bulky, complicated configurations (Mayes et al. 1971, Itoh and Cheng 1972). To meet the requirement of advanced modern mobile communication base station antennas, an optimal combination of a flat dipole antenna and short-circuited patch antenna has been successfully developed, which has been well-known as magnetoelectric dipole antennas (Luk and Wong 2006, Luk and Wu 2012) (Figure 4.1f–g). In this design, one end of two parallel plates is vertically mounted above a metallic reflector, to form a short-circuited patch antenna that serves as a magnetic dipole, with the other end connected to an electric dipole. The antenna can be effectively excited by a coaxial probe in proximity. In these ways, complementary dipole antennas with wideband, single- and dual-polarization characteristics have been implemented (Luk and Wong 2006, Luk and Wu 2012).

Typically, the aforementioned advanced complementary dipole antennas still exhibit a relatively high height of from 0.1- to 0.25-wavelength. It is an interesting and challenging task to design planar complementary dipole antennas with low profiles less than 0.1-wavelength. Thus it leads to all kinds of electrically small Huygens source antenna designs with diverse performance (Lu et al. 2010, Best 2010, Tang et al. 2019): A wideband planar, antipodal Huygens source antenna made up of an open loop and a dipole has been conceptually designed in (Lu et al. 2010). The antipodal dipole is embedded within an open loop to form a unidirectional, planar complementary dipole antenna with a bandwidth up to about one octave. Electrically small, multi-element loop-dipole Huygens source antennas have been theoretically studied (Best 2010). Under multi-port excitations, the magnitude and phase difference between electric and magnetic dipoles can be independently and flexibly controlled, such that a steerable beam can be realized. Multi-layered design with dual-band characteristics has been investigated (Tang et al. 2019). Antenna size can be effectively reduced by incorporating stacked, multi-layered configuration while unidirectional radiation characteristics can be maintained.

In the following sections, we will show how the multi-mode resonant concept can be applied to design different kinds of complementary dipole antennas. At first, we will begin with the multi-mode resonant, planar complementary dipole antennas with a height less than 0.1-wavelength. Then, the multi-mode resonant concept will be employed to simplify the configuration of the complementary antenna while maintaining a wideband characteristic of no less than 30%. In final, a new variety of complementary dipole antenna called *planar self-balanced magnetic dipole antenna* will be introduced: By transforming the unbalanced current into a constructive, resonant electric current sheet, a generalized complementary dipole antenna with a simple planar configuration can be designed.

4.2 MULTI-MODE RESONANT PLANAR, BALANCED COMPLEMENTARY DIPOLE ANTENNAS

Motivated by the development of mobile communications, various dual-band and ultra-wideband complementary dipole antennas have been studied in recent years (Lu, Liu et al. 2014, Wu et al. 2015, Lu, Zhang et al. 2014). In these designs, multiple radiators or relatively complex configurations, e.g., elaborated capacitive compensation network, tapered smooth profiles, etc., need to be introduced (Lu, Liu et al. 2014, Wu et al. 2015, Lu, Zhang et al. 2014). Among them, the multi-mode resonant planar balanced complementary dipole antennas (Lu, Liu et al. 2014) should exhibit the simplest configuration. It can be treated as the combination of a loop and a truncated, meshed traveling-wave dipole. Unlike the traditional single-ended fed, 3-D designs (Wheeler 1947, Kandoian 1951, Clavin 1954, King and Owyang 1960, Mayes et al 1971, Itoh and Cheng 1972, Luk and Wong 2006, Luk and Wu 2012), the antenna exhibits planar, balanced configurations. To validate the antenna's performance, planar baluns are required.

4.2.1 Dual-Band Balanced Loop-Dipole Antennas

A dual-band, balanced loop-dipole antenna can be designed from a classical conical antenna, by incorporating a loop element to a planar, meshed conical dipole. The loop is approximately conformal to the E-field line of the conical dipole, and both elements are co-located excited (Lu, Liu et al. 2014). In this way, a unidirectional, heart-shaped radiation pattern can be yielded (Luk and Wu 2012). In order to achieve dual-band operation, a shorter resonant dipole can be embedded and excited with the loop-dipole combined element simultaneously. The configuration and design

FIGURE 4.2   Photograph of the fabricated dual-band, balanced loop-dipole antenna prototype under test.

process has been illustrated in (Lu, Liu et al. 2014). Figure 4.2 shows a fabricated prototype under test in Satimo's Starlab antenna near the field measurement system. Parametric studies on operation principle and performance optimization have been intensively presented in (Lu, Liu et al. 2014). As can be seen, a wideband transition from broadside coupled stripline to coplanar waveguide (Lu et al. 2011, Lu et al. 2012, Lu et al. 2013) is employed as the balun of the antenna.

To validate its characteristic, prototypes with different balun lengths have been designed and studied. Herein, a prototype antenna fed by a 2 mm shorter balun (Lu, Liu et al. 2014) is investigated, and only the case mounted above a metallic reflector is studied. The simulated and measured frequency responses of the reflection coefficient are illustrated in Figure 4.3. As can be observed, the measured impedance bandwidth for $|S_{11}|$ lower than −10 dB is 73.4% from 1.64 to 3.54 GHz, and 20.7% from 4.90 to 6.03 GHz, which reasonably well matches the simulated ones. Therefore, the antenna can cover most of the land mobile communication spectra from 1.7 to 6 GHz. It is seen the balun's length may slightly affect the reflection coefficient of the antenna (Lu, Liu et al. 2014).

Figure 4.4 shows the simulated and measured radiation patterns at 2.7 and 5.5 GHz, respectively. Good agreement between simulated and measured

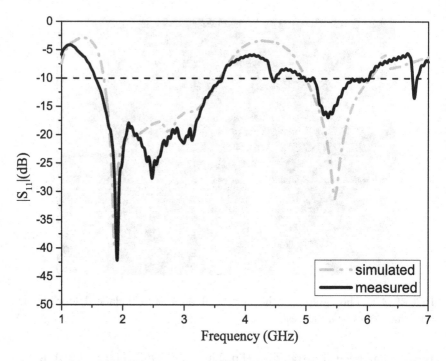

FIGURE 4.3 Simulated and measured reflection coefficient frequency responses of the dual-band, balanced loop-dipole antenna.

patterns can be observed. Within the antenna's impedance bandwidth, the radiation pattern exhibits a stable unidirectional beam with an acceptable cross-polarized level. As can be seen, the radiation patterns are quite similar to the 1.8, 2.4, and 5.3 GHz cases presented in (Lu, Liu et al. 2014), which indicates that the antenna could exhibit relatively stable, wideband radiations within the two respective bands. Since the high-order resonant modes have been sufficiently excited, the side-lobe level will increase, especially in the high-frequency case, while the cross-polarization level within the main beam can be maintained as −15 to −20 dB (5.5 GHz).

The simulated and measured gains are given in Figure 4.5. The measured average gains are quite similar to those presented in (Lu, Liu et al. 2014). The average gains in both bands are 6.8 and 6.5 dBi, respectively, with a discrepancy less than 1.2 dB with the simulated ones. Therefore, it can be determined that the antenna should exhibit less frequency dispersive unidirectional radiation performance at both bands.

As is concluded, the dual-band, planar complementary dipole antenna can satisfactorily cover most of the wireless spectra for land mobile

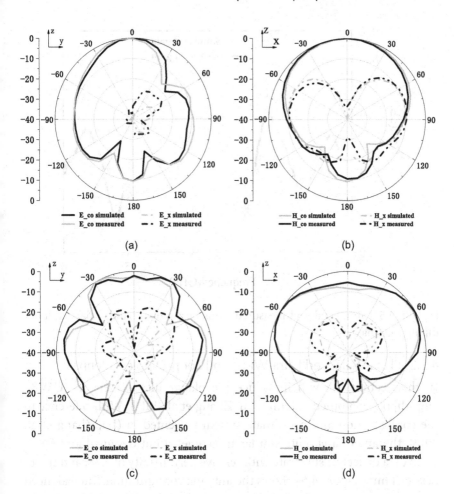

FIGURE 4.4   Simulated and measured E-plane (*yz*-plane) and H-plane (*xz*-plane) radiation patterns of the dual-band, balanced loop-dipole antenna: (a) E-plane, 2.7 GHz, (b) H-plane, 2.7 GHz, (c) E-plane, 5.5 GHz, and (d) H-plane, 5.5 GHz.

communications, such as DCS-1800 (1.71 to 1.88 GHz), WCDMA, and cdma2000 (1.92 to 2.17 GHz), the fifth-generation in China (4.9–5.0 GHz), WLAN (2.4 to 2.484, 5.15 to 5.25, 5.25 to 5.35 and 5.725 to 5.875 GHz) and LTE (2.3 to 2.4, 2.5 to 2.69 GHz). In addition, its low-profile, unbalanced monopole version (Wu et al. 2015) with elaborated configuration has also been validated to exhibit good, dual-wideband performance. Therefore, the conceptual design of loop-dipole antenna (Lu, Liu et al. 2014) and its monopole version (Wu et al. 2015) can be used as the design prototype of multi-band, indoor base station antennas in land mobile communications.

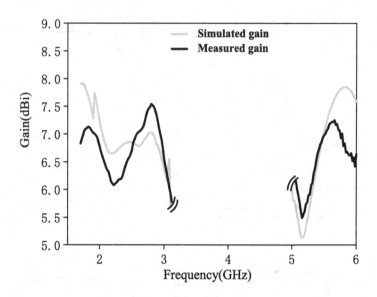

FIGURE 4.5 Simulated and measured gains of the dual-band, balanced loop-dipole antenna.

4.2.2 Planar Balanced Ultra-Wideband Loop-Dipole Antennas

In this section, we will introduce the planar ultra-wideband (UWB), loop-dipole combined antenna (Lu, Zhang et al. 2014). With reference to the configuration and coordinate system presented in (Lu, Zhang et al. 2014), the antenna is quite similar in geometry but designed and fabricated on a dielectric substrate with relative permittivity of 3.38, and thickness of 1 mm. Figure 4.6a shows the antenna configuration. The balanced antenna is composed of a closed, tapered loop and a planar conical dipole. Compared to the case in (Lu, Zhang et al. 2014), the dipole is slightly modified with both corners cut, for better UWB impedance matching. A microstrip line to broadside coupled stripline transition is used to feed the balanced UWB complementary antenna and validate its performance, as illustrated in Figure 4.6b.

It has been validated in (Lu, Zhang et al. 2014) that the antenna's inherent UWB impedance bandwidth would be hardly affected by the externally connected balun. As can be seen from Figure 4.7, the measured impedance bandwidth for $|S_{11}|$ lower than −10 dB is about 135%, from 2.9 to 15 GHz. Compared to the balanced, open-loop case without metallic via-hole connection in (Lu et al. 2010), the impedance bandwidth has been nearly doubled by incorporating a closed-loop configuration.

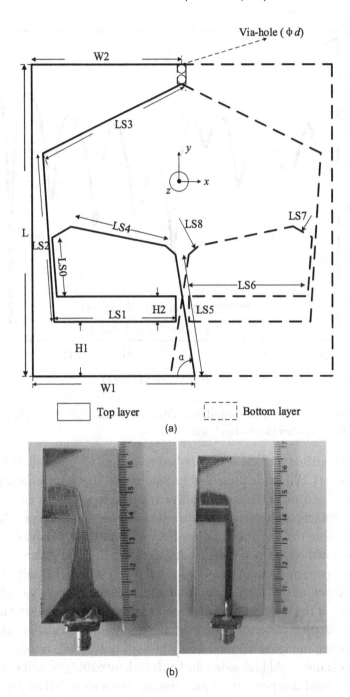

FIGURE 4.6 Planar balanced ultra-wideband loop-dipole antenna: (a) configuration and (b) photograph of some fabricated prototypes.

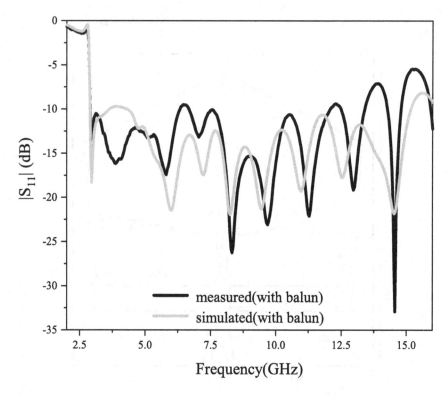

FIGURE 4.7 Measured and simulated reflection coefficients of the planar balanced ultra-wideband loop-dipole antenna.

The simulated and measured radiation patterns in E- and H-planes at 3.0 and 11.2 GHz are presented in Figure 4.8a to 4.8d. Generally, the simulated and measured results agree reasonably well with each other with acceptable discrepancies. As can be observed from Figure 4.8a and 4.8b, it is seen that the antenna's low-frequency radiation behavior should be dominated by the loop radiator. As operating frequency increases, the radiation pattern should be co-dominated by the simultaneously excited electric dipole as well as the loop. This implies that the antenna would tend to operate at the Huygens-source mode, thus unidirectional, nearly heart-shaped radiation patterns could be attained in both principal-cut planes, as illustrated in Figure 4.8c and 4.8d. The high-order mode resonant loop and dipole may yield high side-lobe levels (Schantz 2015). In addition, the skewed E-field components on the substrate cross section (Langley et al. 1993) may cause cross-polarized components, thus it leads to degradation of polarization purity. Nevertheless, the antenna can exhibit a relatively

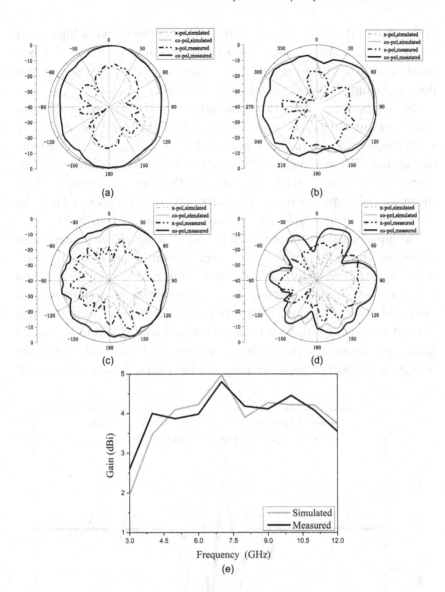

FIGURE 4.8   Simulated and measured radiation patterns and gain of the planar balanced ultra-wideband loop-dipole antenna: (a) *yz*-plane, 3.0 GHz, (b) *xy*-plane, 3.0 GHz, (c) *yz*-plane, 11.2 GHz, (d) *xy*-plane, 11.2 GHz, and (e) antenna gain.

stable, +*y*-oriented, endfire main beam over the whole impedance bandwidth. The measured peak gain of the proposed antenna is presented in Figure 4.8e, and it is reasonably matched to the simulated one. The in-band average measured peak gain is about 4.0 dBi, which is comparable to the case in (Lu, Zhang et al. 2014). Due to the excitation of high-order resonant modes, the radiation pattern may be degraded by the side lobes. In addition to the parasitic radiations of the balun, the antenna gain may drop at the high-frequency band (Wu et al. 2015).

For further time-domain characteristic validation, the endfire-to-endfire transfer function (magnitude and group delay) measurement setup has been shown in (Lu, Zhang et al. 2014). The separation between transmitted (Tx) and received (Rx) antennas is $D = 100$ cm. The test is performed in an indoor environment filled with arbitrary scatterers. Then, the measured frequency-domain transfer function is transformed into the time-domain one by employing the inverse fast Fourier transform.

The input signal for transmission is selected to be the fourth derivative of Gaussian pulse that satisfies the Federal Communications Commission (FCC) spectral mask for indoor UWB applications. The pulse fidelity function (Quintero et al. 2011) of ρ is used to evaluate the performance, as discussed in (Lu, Zhang et al. 2014). In the calculation of ρ, both transmitted and received signals have been normalized and plotted in Figure 4.9 for comparisons. It is seen that the pulses agree fairly well with each other. Ripples and late-time signal ringing can be observed, which are caused

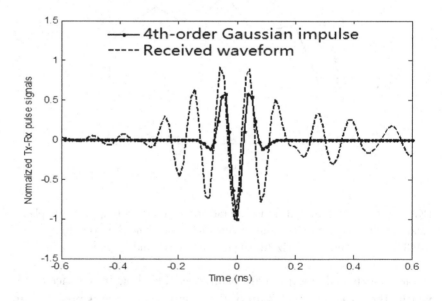

FIGURE 4.9 Pulse signal fidelity of the planar balanced ultra-wideband loop-dipole antenna for receiving fourth-order Gaussian pulse signal.

by the low-frequency spectra leakage and the antenna system's time dispersion (including the balun). A more compact antenna size may be beneficial to mitigate such dispersions and signal ringing (Schantz 2015), thus a lumped, miniaturized balun would be a better choice for practical impulse signal transmitting and receiving. The calculated pulse fidelity is $\rho = 0.6707$, which implies that the signal dispersion would be acceptable (Lu, Zhang et al. 2014) in the transmission of UWB impulse signals. As discussed in (Lu, Zhang et al. 2014), the antenna would exhibit better pulse fidelity when higher-order derivative Gaussian signals were transmitted.

In this section, a planar UWB, balanced loop-dipole antenna with endfire radiation has been investigated. The operation bandwidth of the antenna is up to 135% for $|S_{11}|$ lower than $-10\,\mathrm{dB}$. Compared to most of the existed complementary dipole antennas, the advanced antenna has merits of compact size (i.e., $0.23\lambda_{\mathrm{Lg}} \times 0.23\lambda_{\mathrm{Lg}}$, where λ_{Lg} denotes the lowest cut-off guided-wave wavelength), and simple balanced configuration. In addition, the proposed antenna exhibits acceptable pulse signal fidelity. Therefore, it would be a promising antenna design for all kinds of compact, portable UWB wireless devices.

4.2.3 Dual-Mode Resonant Planar Endfire Circularly Polarized Antennas

The complementary dipole antennas discussed in the previous sections are linearly polarized (LP) antennas. In LP complementary dipole antennas designs, the equivalent magnetic dipole should be orthogonal to the electric dipole so that they could share the identical polarization and lead to a unidirectional LP characteristic. When the magnetic dipole is incorporated in parallel to the electric dipole, with both dipoles properly excited with equal magnitude and phase quadrature, CP radiation would be attained instead (Kandoian 1951).

Planar endfire circularly polarized antenna (PECPA) is a kind of low-profile complementary dipole antennas. It can generate CP, endfire beam in parallel with its plane. Basically, PECPAs can be made up of orthogonal magnetic dipoles (Lu et al. 2015) or complementary dipoles (Zhang et al. 2016, Xue et al. 2016, You et al. 2016, Yang et al. 2018, Li et al. 2016, Zhang et al. 2016, Lu et al. 2019), with design guidelines illustrated in (Lu et al. 2015, Zhang et al. 2016, Xue et al. 2016, You et al. 2016, Yang et al. 2018, Li et al. 2016, Zhang et al. 2016). Generally, PECPAs and their variant, bi-sensing counterpart (Chen et al. 2019) exhibit relatively narrow impedance bandwidth (Xue et al. 2016, You et al. 2016, Yang et al. 2018) of less than 5%. V-shaped open loop configuration with partial traveling-wave effect (Wheeler 1950, Xu et al. 2019) can yield wideband, broadened beamwidth designs (You et al. 2016, Yang et al. 2018) with an enhanced front-to-back ratio (FBR). In these designs, the axial ratio (AR) frequency response still exhibits a narrowband, single-mode resonant characteristic of no more than 11% (You et al. 2016, Yang et al. 2018). Thus, the multi-mode resonant concept can be introduced to enhance both the impedance and AR bandwidths (Li et al. 2016, Zhang et al. 2016).

Figure 4.10a shows a conceptual design of PECPA (Li et al. 2016) with dual-mode resonant reflection coefficient and AR frequency responses. It is composed of a rectangular magnetic dipole, and a unilaterally loaded, straight dipole, with design parameters fully given in (Li et al. 2016). As shown in Figure 4.10b, two resonances at 2.45 GHz and 2.53 GHz can be obtained when the stub length is set as 22.3 mm. The impedance band-width for $|S_{11}|$ lower than −10dB of the stub-loaded, dual-mode reso-nant antenna is from 2.40 to 2.63 GHz, which is 9.4% in fraction, and much wider than the 2% ones presented in (Lu et al. 2015) and (Zhang et al. 2016). As is seen from the measured and simulated AR frequency responses in Figure 4.10b, the antenna can exhibit a dual-mode resonant AR frequency response. At the center band, the AR drastically increases to over 6 dB, which is poorly matched to the simulated one. Such discrepancy may be caused by fabrication errors, and it can be mitigated by carefully tuning the position and length of the loaded stub. Nevertheless, the dual-mode resonant phenomenon is unchanged by the discrepancy. This indi-cates that a stub-loaded, dual-mode resonant dipole may offer both wider impedance and AR bandwidth to PECPAs.

The conceptual design illustrated in Figure 4.10a exhibits a relatively bulky size, as a truncated, rectangular leaky-wave microstrip antenna is used as the magnetic dipole. For more compact and simpler designs, reso-nant concentric annular configuration (Xue et al. 2016) can be employed to serve as a promising prototype without changing the operational prin-ciple. Figure 4.10c and d show the configuration of dual-mode resonant PECPA, and a fabricated prototype on foam substrate (Zhang et al. 2016), respectively. It is composed of a semicircular magnetic dipole, and a con-centric annular sector, stub-loaded dipole: The semicircular magnetic dipole serves as a vertically polarized radiator, while the concentric annu-lar sector dipole acts as an x-polarized horizontal radiator. In this way, it

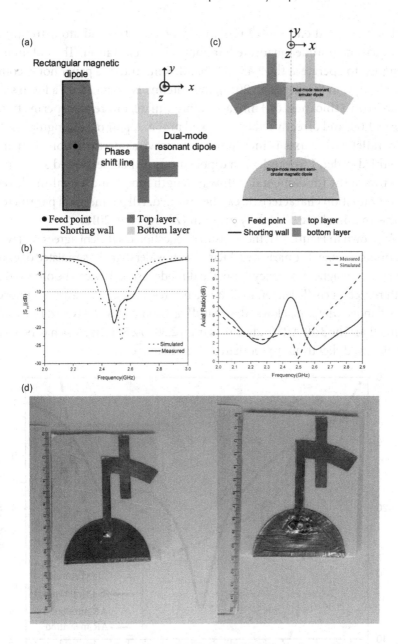

FIGURE 4.10   Dual-mode resonant PECPAs: (a) PECPA made up of a rectangular magnetic dipole and dual-mode resonant straight dipole, (b) simulated and measured impedance and AR bandwidths, (c) PECPA composed by a dual-mode resonant annual dipole and a single-mode resonant semicircular magnetic dipole, and (d) photograph of a fabricated prototype on foam substrate.

yields *y*-oriented endfire CP radiation when the two radiators are under an equal magnitude, quadrate temporal phase excitation. The antenna is designed to operate at the 2.45 GHz band. Similar to the paper honeycomb counterpart in (Zhang et al. 2016), the antenna is designed on a foam substrate with a thickness of 4 mm and approximate, low relative permittivity of ε_r = 1.02, and directly fed by a coaxial probe. A pair of rectangular stubs in parallel with *y*-axis is incorporated to serve as perturbation elements. The third-order, 1.5-wavelength dipole mode is then perturbed and tuned downward the fundamental, half-wavelength mode, and a dual-mode resonant radiation characteristic can be attained. All geometrical parameters of the antenna have been fully given in (Zhang et al. 2016).

As shown in Figure 4.11, the measured reflection coefficient agrees well with the simulated one. The measured AR is slightly wider than the simulated one and deviates to a higher frequency band. Dual-mode resonances can be observed in both reflection coefficient and AR frequency responses. The measured impedance bandwidth for $|S_{11}|$ lower than −10 dB is from 2.21 to 2.84 GHz, i.e., 23.6% in fraction. The 3 dB AR bandwidth is from 2.30 to 2.80 GHz, which is 19.6% in fraction, and slightly narrower than the impedance one.

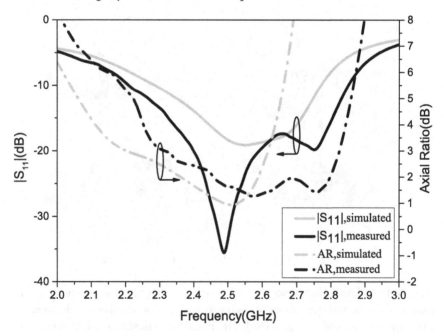

FIGURE 4.11 Simulated and measured reflection coefficients and ARs of the dual-mode resonant PECPA.

The simulated and measured radiation patterns in *xy*- and *yz*-planes, at 2.50 GHz are plotted in Figure 4.12a and b. As can be seen, the antenna exhibits a +*y*-oriented, endfire CP radiation pattern: The 3 dB AR beamwidths in azimuth and elevation planes are 60° and 90°, respectively. Figure 4.13 plots the simulated and measured peak gains. As can be seen, the

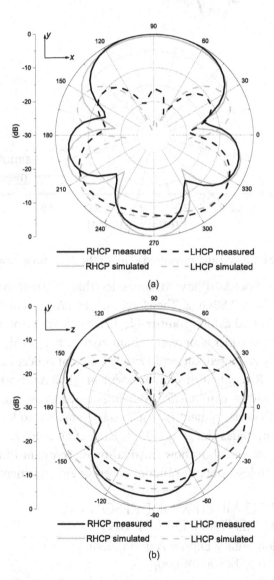

FIGURE 4.12 Radiation patterns of the dual-mode resonant PECPA at 2.5 GHz: (a) azimuth pattern and (b) elevational pattern.

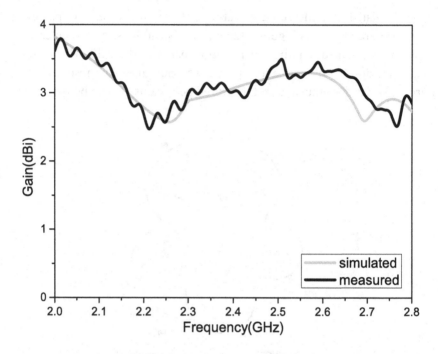

FIGURE 4.13 Simulated and measured gains of the dual-mode resonant PECPA.

average peak is about 3.0 dBic with flatness less than 1 dB within the AR band-width from 2.30 to 2.80 GHz. Therefore, the PECPA implemented on foam substrate can indeed exhibit comparable dual-mode resonant, wideband CP performance to its paper honeycomb counterpart (Zhang et al. 2016).

As evidently validated, the multi-mode resonant concept can be indeed effective for AR bandwidth enhancement of PECPAs. Both impedance and AR bandwidths can be effectively broadened to over or nearly 20%. The bandwidth enhancement is remarkable compared to its single-mode resonant counterparts (Lu et al. 2015, Zhang et al. 2016, Xue et al. 2016). In Chapter 6, we will show how the dual-mode resonant PECAPs can be applied in a wireless router for indoor coverage enhancement.

4.3 SIMPLIFIED MULTI-MODE RESONANT COMPLEMENTARY DIPOLE ANTENNAS

4.3.1 Complementary Dipole Antenna Using Even-Mode Resonant Loop

The multi-mode resonant concept can be employed to further simplify the structural complexity of complementary dipole antennas while maintaining a wideband unidirectional performance simultaneously: When a square loop is excited at its center via a short dipole, rather than the diagonally fed cases in (Yuan et al. 2018, Xu et al. 2019), even-mode loop antenna with LP characteristic can be attained. If the upper and bottom zones of the loop are extended to form a small "ground plane", and to facilitate the bonded coaxial feed line, hybrid resonances of loop and slot mode can be yielded, as depicted in Section 3.5. Then, the hybrid-mode resonant loop can be further combined with a dipole via a quarter-wavelength transmission line. In these ways, it yields a simple unidirectional, hybrid-mode resonant, loop-dipole combined antenna with 12 design parameters (Chen et al. 2017) to provide an available bandwidth of up to 30%.

The loop-dipole combined antenna can be approximately analyzed and designed by using the equivalent sources model (Chen et al. 2017). Figure 4.14 shows the antenna and its fabricated prototype. For wideband impedance matching, the square loop has been modified into a rectangular one with a length-to-width ratio of 0.78, where the perimeter is increased to about 1.22-wavelength. The loop element is fabricated on a substrate with relative permittivity of $\varepsilon_r = 2.65$, tan $\delta = 0.002$ and thickness $h = 1.0$ mm. As shown in the photograph of Figure 4.14b, an adhesive coaxial cable bonded at the loop's center serves as the antenna feed. The design parameters have been fully given in (Chen et al. 2017).

The simulated and measured results of reflection coefficient frequency responses are illustrated in Figure 4.15. As can be seen, the measured impedance bandwidth for $|S_{11}|$ lower than -10 dB ranges from 2.25 to 3.05 GHz (i.e., 30.2% in fraction), which matches reasonably well with the simulated one. Owning to the relatively small width-to-length ratio of the rectangular loop and the relatively large "ground plane" surrounding both upper and bottom zones of the loop, the "slot mode" discussed in Chapter 3 could hardly be fully suppressed. This may yield the hybrid resonances effect as depicted in Chapter 3, and a parasitic resonance could be observed near 2.5 GHz. Thus the presented antenna can be approximately recognized as a hybrid, multi-mode resonant design. The normalized simulated and measured radiation patterns at 2.46 and 2.89 GHz are depicted in Figure 4.16. At all frequencies, unidirectional radiation patterns with FBR higher than 20 dB can be attained. The cross-polarization level within the main beam is lower than -18 dB. The good agreement between simulated and measured results indicates that the presented design approach should be effective for unidirectional antenna designs without the aid of a metallic reflector.

Figure 4.17 shows the simulated and measured bore-sight gains. Both simulated and measured curves exhibit a similar trend. The measured

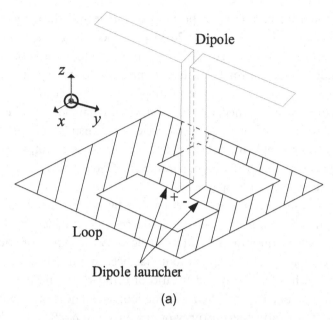

(a)

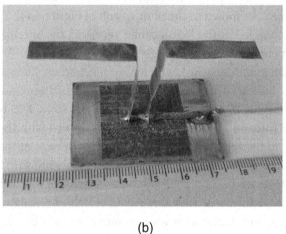

(b)

FIGURE 4.14   Complementary dipole antenna using even-mode resonant loop: (a) configuration and (b) photograph of a fabricated prototype.

in-band average gain is 5.0 dBi, which is slightly dropped than the simulated one of 5.5 dBi with a maximum discrepancy of less than 1.0 dB, which is acceptable and reasonable. It should be caused by insufficiently suppressed hot-cable effect, and the unpredicted dielectric and conductor losses.

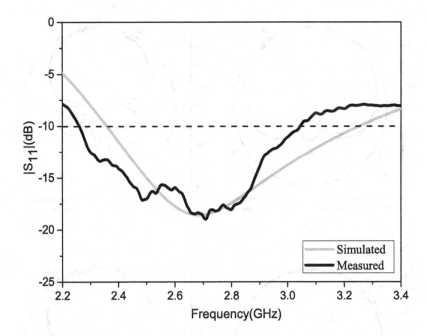

FIGURE 4.15 Measured and simulated reflection coefficients of the complementary dipole antenna using an even-mode resonant loop.

As is illustrated in Figures 4.15–4.17, it is seen that a complementary dipole antenna using even-order mode resonant loop element under hybrid loop-slot modes resonance should exhibit an available radiation bandwidth of about 30%. Furthermore, it exhibits a simpler configuration than most of the presented designs (King and Owyang 1960, Mayes et al. 1971, Itoh and Cheng 1972, Luk and Wong 2006, Luk and Wu 2012, Jin and Ziolkowski 2010, Niemi et al. 2012).

4.3.2 Complementary Dipole Antenna Using Dual-Mode Resonant Slotline Element

Typically, complementary dipole antennas with simpler configuration, fewer design parameters, and more compact size (Jin and Ziolkowski 2010, Niemi et al. 2012) can only exhibit available bandwidth of less than 5%. How to design an antenna with simple configuration and remarkably enhanced bandwidth remains a challenge. In this section, a dual-mode resonant slotline antenna operating at its fundamental and 1.5-wavelength slot modes is combined with a straight dipole, which leads to a dual-mode resonant complementary dipole antenna with nine design parameters and available bandwidth over

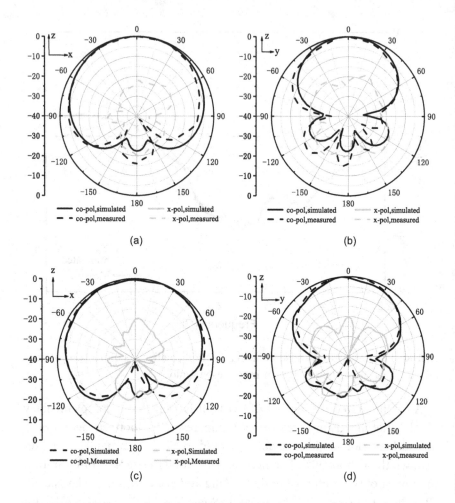

FIGURE 4.16 Measured and simulated radiation patterns of the complementary dipole antenna using hybrid mode resonant loop: (a) xz-plane at 2.46 GHz, (b) zy-plane at 2.46 GHz, (c) xz-plane at 2.89 GHz, and (d) zy-plane at 2.89 GHz.

33% (Cheng et al. 2018). As will be seen, the multi-mode resonant concept is indeed beneficial for simplifying the configuration of complementary dipole antennas, while maintaining their wideband characteristic.

The proposed antenna is shown in Figure 4.18. It is composed of a center-fed, dual-mode resonant slotline radiator and a symmetric, half-wavelength dipole. The dipole is connected to the center of the slotline radiator via a quarter-wavelength, linearly tapered broadside coupled stripline. The design parameters of the antenna have been presented in (Cheng et al. 2018).

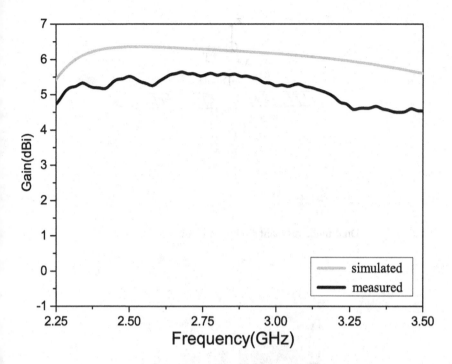

FIGURE 4.17   Measured and simulated bore-sight gains in +z-direction ($\theta = 0°$) of the complementary dipole antenna using a hybrid-mode resonant loop.

The simulated and measured reflection coefficients of the antenna are presented in Figure 4.19. As can be seen, both measured and simulated results are in good agreement and the measured impedance bandwidth for $|S_{11}|$ lower than −10 dB ranges from 1.95 to 2.93 GHz, i.e., 40.2% in fraction. The dual-mode resonant characteristic is contributed by the 1.5-wavelength slotline radiator. Compared to the PECPA using dual-mode resonant electric dipole (Zhang et al. 2016), it can be seen that a dual-mode resonant slotline radiator may incorporate wideband characteristics to the complementary dipole antenna.

The surface E-field distribution within the slotline radiator is simulated and displayed in Figure 4.20. It is seen that the E-field distribution at 2.05 GHz is quite similar to those of fundamental, half-wavelength mode. At 2.65 GHz, the E-field distribution approximately exhibits three maxima, with one's direction being "up" and that of the other two "down", just like the 1.5-wavelength resonant mode behaves. As is indicated, the

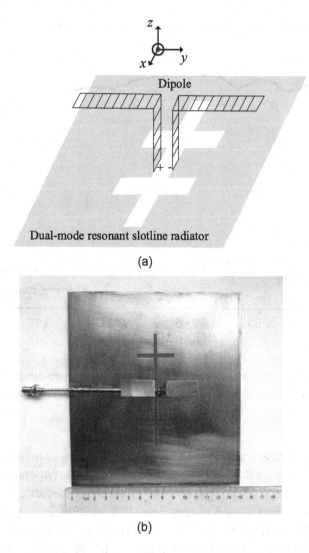

(a)

(b)

FIGURE 4.18   Dual-mode resonant complementary dipole antenna: (a) configuration and (b) photograph of a fabricated prototype.

1.5-wavelength slotline mode has been evidently excited for radiation under the perturbation of slot stubs.

The simulated and measured normalized radiation patterns in both principal-cut, elevation planes at 1.98 and 2.61 GHz are plotted in Figure 4.21. As can be seen, the antenna exhibits stable, unidirectional radiation patterns. The cross-polarization level within both elevation planes is lower than −15 dB within the main beam, which implies that polarization purity should

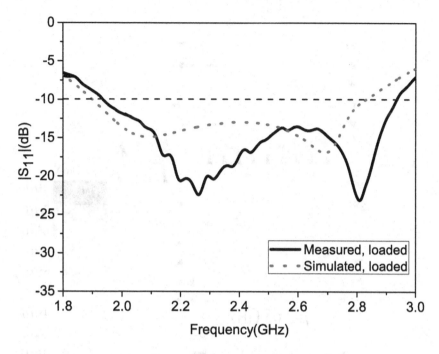

FIGURE 4.19 Measured and simulated reflection coefficients of the complementary dipole antenna using dual-mode resonant slotline radiator.

be acceptable. As indicated in (Cheng et al. 2018), the boresight gain variation at both low and high-frequency bands can reach as high as 2 dB, and the available radiation bandwidth may be degraded. Such a large discrepancy could possibly contribute to two factors: At low-frequency band, it should be caused by the "hot-cable effect" of the feed cable, and at high-frequency band, it should be caused by the sufficiently excited high-order slotline mode. As shown in Figure 4.22, the measured gain within 2.0–2.7 GHz matches well with the simulated one, with a discrepancy smaller than 1 dB. Thus, the available gain bandwidth can be determined as 2.0–2.7 GHz, which is about 30% in fraction and narrower than the impedance one.

As can be seen from this section, it is convinced that the concept of multi-mode resonance can indeed be employed to reduce the complementary dipole antenna's configuration complexity while maintaining its wideband characteristic. Both examples exhibit simple antenna configurations with a few design parameters, while their available radiation bandwidth can be maintained as over 30%.

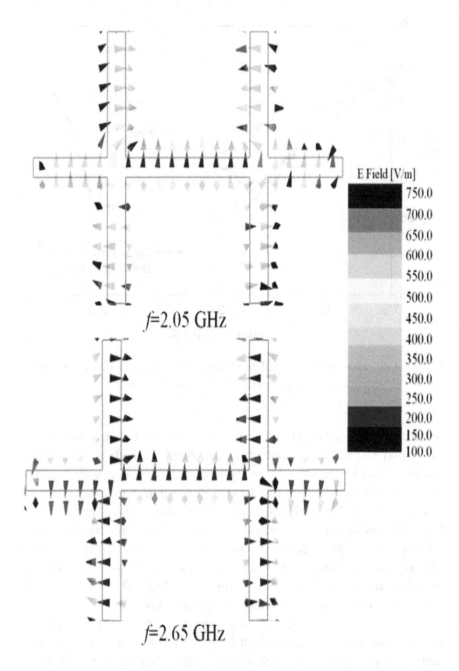

FIGURE 4.20 Simulated E-field distributions of the complementary dipole antenna using dual-mode resonant slotline radiator at 2.05 GHz and 2.65 GHz.

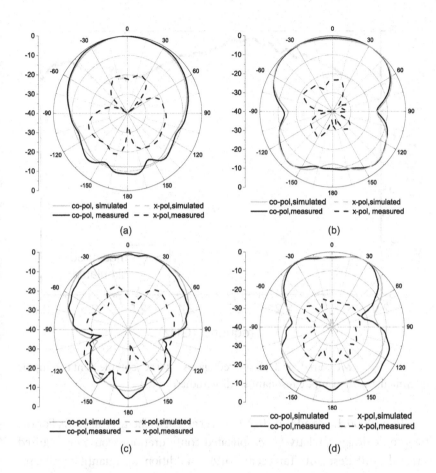

FIGURE 4.21 Measured and simulated normalized radiation patterns of the complementary dipole antenna using dual-mode resonant slotline radiator: (a) xz-plane, at 1.98 GHz, (b) yz-plane, at 1.98 GHz, (c) xz-plane, at 2.61 GHz, and (d) yz-plane, at 2.61 GHz.

4.4 SELF-BALANCED MAGNETIC DIPOLE ANTENNA: NOVEL GENERALIZED, LOW PROFILE MULTI-MODE RESONANT COMPLEMENTARY DIPOLE ANTENNA

4.4.1 Conceptual Design of Planar Self-Balanced Magnetic Dipole Antenna

In the previous sections, wideband complementary dipole antennas have been designed based on a multi-mode resonant electric (Zhang et al. 2016) or magnetic dipole (Chen et al. 2017, Cheng et al. 2018). It is also found that most complementary dipole antennas should exhibit relatively high profiles (Luk

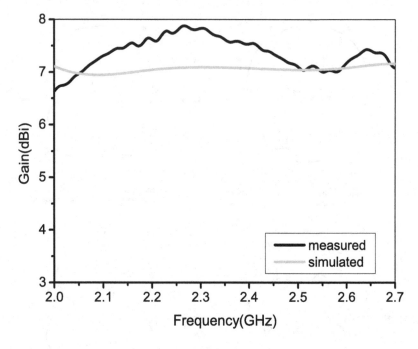

FIGURE 4.22 Measured and simulated gains of the complementary dipole antenna using dual-mode resonant slotline radiator.

and Wong 2006, Luk and Wu 2012, Chen et al. 2017, Cheng et al. 2018). For the low-profile designs, relatively complicated configurations would be required (Lu et al. 2010, Best 2010, Tang et al. 2019). In addition, the quantitative beam-width controlling technique for complementary dipole antennas has been rarely developed. Therefore, it would be always a challenging task to advance effective design approaches to planar complementary dipole antenna with low configurational complexity, and beamwidth manipulating functionality.

Motivated by the challenging task, a new conceptual design approach to planar complementary dipole antennas should be advanced based on a wide beamwidth antenna. Therefore, a quasi-isotropic antenna with widened beamwidth in all principal-cut planes (Li et al. 2017) may serve as a suitable candidate in the initial step. As is well known, the non-uniformity of quasi-isotropic magnetic dipole severely degrades to about 5.6 dB due to the spurious radiations from the feed cable (Li et al. 2017). Since the unbalanced current on the sheath of the cable could not be fully suppressed, it would be a better way to transform such "negatively undesired" currents into the "positively useful" ones: In a more generalized sense, if

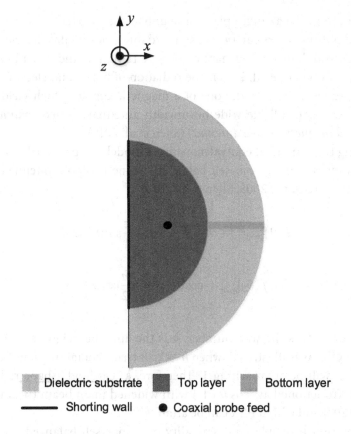

Dielectric substrate   Top layer   Bottom layer
——— Shorting wall   ● Coaxial probe feed

FIGURE 4.23   Planar self-balanced magnetic dipole antenna printed on a finite dielectric substrate.

the unbalanced electric current flowing on the surface of an elementary magnetic dipole antenna can be transformed and employed to contribute constructive radiations, generalized multi-mode resonant, complementary dipole antenna under both electric and magnetic current resonances could be yielded. In this way, the unbalanced effect would be automatically suppressed; meanwhile, the concept of a "self-balanced magnetic dipole antenna" would be attained. Therefore, the evolution process from quasi-isotropic magnetic dipole antenna (Li et al. 2017) to planar self-balanced magnetic dipole antenna (Shen et al. 2019) has been illustrated in a step-by-step manner. As has been revealed in (Shen et al. 2019), a horizontal quarter-wavelength stub can be incorporated to one side of the magnetic dipole. In that way, a virtual magnetic wall boundary condition would be introduced in the azimuth plane. The unbalanced current sheets on both top and bottom layers can be transformed into a horizontal electric dipole that can simultaneously generate effective radiation and a self-balanced effect (Thumvichit et al. 2007). The radiation of horizontal electric dipole can be superposition to the one of a magnetic dipole, which yields unidirectional, controllable wide beamwidth antennas named "*planar self-balanced magnetic dipole antenna*" (Shen et al. 2019).

Using the simplified equivalent sources model, the principal-cut radiation patterns of the planar self-balanced magnetic dipole antenna can be depicted by (Balanis 2005, Shen et al. 2019)

$$f(\theta)\big|_{\phi=0} = 1 + \cos\left(k\frac{h}{2}\cos\theta - \frac{\pi}{4}\sin\theta\right)\cos\theta \tag{4.1}$$

$$f(\theta)\big|_{\phi=90°} = \cos\theta + \cos\left(k\frac{h}{2}\cos\theta\right) \tag{4.2}$$

where $k = 2\pi/\lambda$ is the wavenumber, h is the antenna height, and λ is the wavelength. As is illustrated, when $h\to0$, all terms containing h in Eq. (4.1) and (4.2) will be approaching 1. Therefore, a "cardioid" shape radiation pattern (functioned as "$\cos\theta + 1$") with widened main beam (beamwidth up to 120°) and a high FBR can be attained.

Operation principle and generality of planar self-balanced magnetic dipole antenna have been validated on air substrate by investigating prototype antennas having different flared angles (Shen et al. 2019). Herein, we will design and study a printed version with a lower profile to further validate the effectiveness and generality of the design approach.

Figure 4.24 shows the photograph of a fabricated planar self-balanced magnetic dipole antenna prototype. The antenna is designed at 2.4 GHz, and printed on dielectric substrate with relative permittivity $\varepsilon_r = 2.65$, and height $h = 2$ mm. A quarter-wavelength stub is incorporated to the bottom layer of the semicircular magnetic dipole to form a planar self-balanced magnetic dipole antenna with an extremely low profile of about 0.016-wavelength (in free space wavelength).

Figure 4.25 shows the simulated and measured reflection coefficients of the fabricated prototype. As can be seen, the measured center frequency perfectly agrees with the simulated one. The measured impedance bandwidth for $|S_{11}|$ lower than -10 dB is from 2.32 to 2.54 GHz, which is about

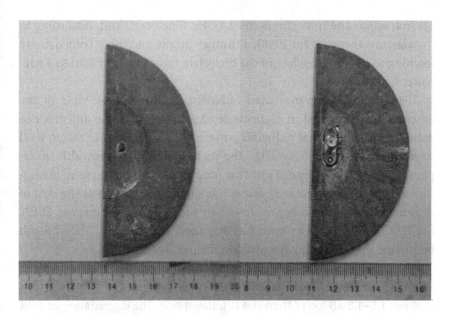

FIGURE 4.24 Photograph of a fabricated planar self-balanced magnetic dipole antenna printed on a dielectric substrate.

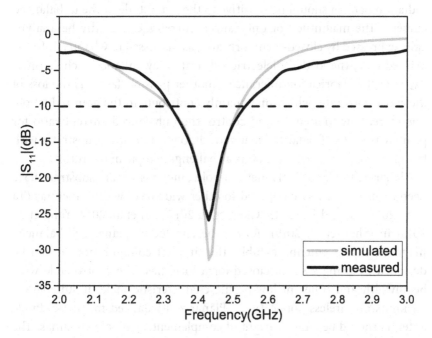

FIGURE 4.25 Simulated and measured reflection coefficients of the printed antenna prototypes.

5% in fraction and fully covers the 2.4 GHz Bluetooth band. According to the classical theory (Chu 1948), a thinner profile and more compact size (owning to the loaded effect of the dielectric substrate) may lead to a narrower bandwidth.

The simulated and measured radiation patterns at 2.45 GHz of the antenna are illustrated in Figure 4.26. As can be seen, the antenna can exhibit a unidirectional radiation pattern with a cross-polarization level lower than −10 dB at 2.45 GHz. The measured main beamwidth of xz-plane is 95°, while the one of yz-plane is up to 180°. As can be seen, the significantly widened yz-plane beamwidth should be attained at the cost of degradation of FBR: Compared to the air substrate case (Shen et al. 2019), the FBR of the printed antenna has degraded to about 6 dB. The FBR can be further improved and discussed in the next section.

The simulated and measured radiation gains in the bore-sight, +z-direction are plotted in Figure 4.27. As can be seen, the measured gain is about 1.5–1.8 dB lower than the simulated one. The degradation of FBR and gain decrement may be caused by two possible reasons: One is the sensitive unbalanced current ratio. As formulated in (Shen et al. 2019), the radiation pattern should be sensitive to the magnitude of the unbalanced current. The magnitude of unbalanced currents can hardly be analytically expressed by closed-form formulas, as discussed in Chapter 1. In the printed case, the loaded dielectric substrate may enhance such sensitivity, as well as fabrication tolerance. Another possible factor is the loss of dielectric substrate, which may mainly contribute to the gain reduction. Therefore, how to analyze and control the unbalanced current ratio for printed, planar self-balanced magnetic dipole antennas on substrates with high relative permittivity remains a challenging topic in the future.

Planar self-balanced magnetic dipole antennas on air substrate have been comprehensively compared to other wide beamwidth antennas (Ta et al. 2013, Ko and Lee 2013, Chen et al. 2017, Pan et al. 2014, Yang et al. 2018) in (Shen et al. 2019). As can be concluded, planar self-balanced magnetic dipole antenna exhibits the simplest configuration. It can be designed by using a set of closed-form formulas. The controllable wide beamwidth and inherent low-profile characteristic make it suitable for all kinds of wireless applications. Planar self-balanced magnetic dipole antenna should be a new variant of complementary dipole antennas: The unbalanced currents on the outer surface of the magnetic dipole can be properly transformed by incorporating a parasitic element, such that the

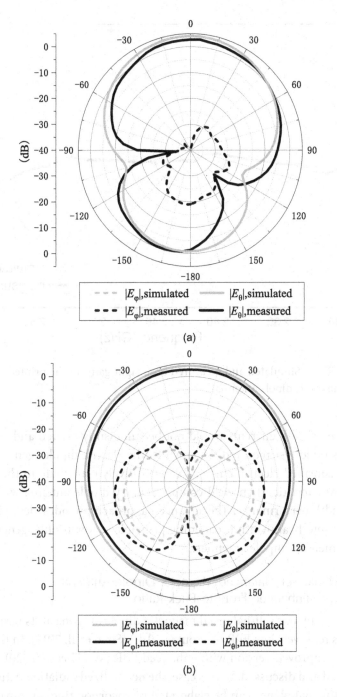

FIGURE 4.26 Simulated and measured radiation patterns of the printed self-balanced magnetic dipole antenna at 2.45 GHz: (a) *xz*-plane and (b) *zy*-plane.

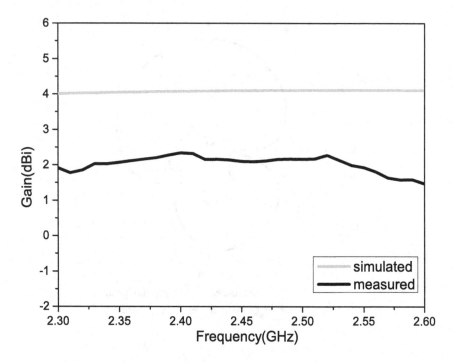

FIGURE 4.27   Simulated and measured bore-sight gains of the printed self-balanced magnetic dipole antenna.

undesired sheath currents would not be sufficiently excited and external balun is no longer required. Meanwhile, the electric dipoles generated by the transformed electric currents can positively contribute to the radiation characteristic, then yield complementary dipole antenna with wide beamwidth performance. The advanced approach provides clear physical insights into beamwidth broadening principle of low-profile, generalized complementary dipole antennas.

4.4.2 Planar Self-Balanced Magnetic Dipole Antenna with Enhanced Front-to-Back Ratio

As is seen, planar self-balanced magnetic dipole antenna at its basic form exhibits relatively low FBR of about 6–10 dB (Shen et al. 2019). In this section, an improved version with enhanced FBR (Wu H et al. 2020) will be presented and discussed. In this case, the normalized radiation patterns in both principal planes can be elaborated by incorporating an unbalanced current ratio (UCR) indicated as b

$$f(\theta)|\phi=0 = (1-b) + 2b\cos\left(\frac{kh}{2}\cos\theta - \frac{\pi}{4}\sin\theta\right)\cos\theta \qquad (4.3)$$

$$f(\theta)|\phi=90^\circ = (1-b)\cos\theta + 2b\cos\left(\frac{kh}{2}\cos\theta\right) \qquad (4.4)$$

As can be seen and validated in (Wu H et al 2020), the radiation patterns are functioned by the UCR of b. Actually, due to the complexity of the connection effect between the antenna and feed, it is hard to deduce a brief, closed-form formula for the unbalanced current ratio, as previously formulated in Chapter 1. In practice, the regulation and control of the UCR factor of b can be implemented by modifying the parasitic stub's shape (Wu H et al. 2020). Herein, we will present an alternative design quite similar to the Ant. 6 in (Wu H et al. 2020). Most of the parameters of the presented antenna are identical to those of Ant. 6, with the end arc's angle α slightly modified to 4.5° and an additional rectangle inserted between another end of the stub and the circumference of the magnetic dipole.

Figure 4.28 shows the photograph of a fabricated prototype. As can be seen, the bottom layer of the magnetic dipole has been slightly extended

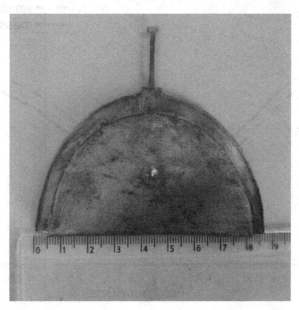

FIGURE 4.28 Planar self-balanced magnetic dipole antenna with a modified parasitic element.

with an extruded rectangle, and the parasitic stub has been modified into an arc-loaded configuration. To finely tune the UCR factor of b, an additional rectangle with 7 mm in length, and 1.5 mm in width is incorporated between the stub and the circumference of the magnetic dipole. The simulated and measured reflection coefficients of the antennas are plotted in Figure 4.29. As is shown, the simulated result matches well with the measured one, with a discrepancy of center frequency less than 1%. The antenna achieves an impedance bandwidth for $|S_{11}|$ lower than −10 dB from 2.24 to 2.54 GHz (12.6%).

The normalized radiation patterns of the antenna are measured at 2.42 GHz and compared with the simulated ones. As shown in Figure 4.30, the measured patterns match reasonably well with the simulated ones. The measured FBR is 20 dB in −z-direction ($\theta = 180°$). As observed in the xz-plane pattern, two nulls can be found in the back lobe, which occurring in $\theta = 135°$ and $\theta = 165°$ directions, respectively. This implies the FBR of planar self-balanced magnetic dipole antenna should be quite sensitive to the UCR of b. If the null can be steered to $\theta = 180°$ by properly tuning the

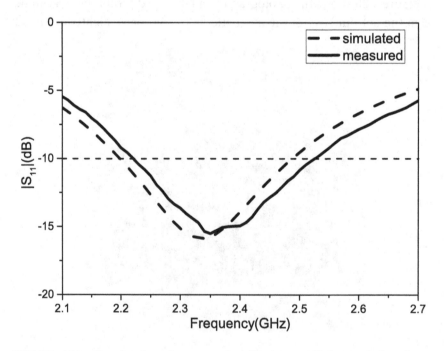

FIGURE 4.29 Simulated and measured reflection coefficients of the planar self-balanced magnetic dipole antenna with a modified parasitic element.

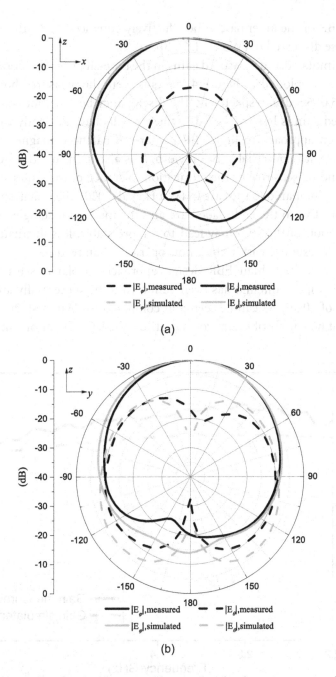

FIGURE 4.30 Simulated and measured normalized radiation patterns of the planar self-balanced magnetic dipole antenna with a modified parasitic element at 2.42 GHz: (a) xz-plane and (b) zy-plane.

UCR, FBR of the antenna can be effectively enhanced to 30 dB or more, like those discussed in (Wu et al. 2020).

The simulated and measured gains in the bore-sight, +z-direction of the antennas are plotted in Figure 4.31. As is seen, the measured bore-sight gain is 5.0–5.5 dBi, respectively. The discrepancy between measured and simulated gain is less than 0.6 dB. These results have evidently validated the design approach to high FBR planar self-balanced magnetic dipole antennas with controllable beamwidth. It is also seen that the UCR factor should be sensitive to the fabrication tolerance of the stub's shape in practice: Compared to the case in (Wu H et al. 2020), the radiation gain is quite sensitive to the end-loaded arc and the inserted rectangle. A slight modification of the stub may lead to a more stable, less fluctuated gain frequency response within the whole operation bandwidth.

In this section, high FBR design approach to planar self-balanced magnetic dipole antenna has been validated to be successfully attaining an FBR of 20 dB. As can be seen and compared to (Wu et al. 2020), it is seen that how to flexibly control and utilize the UCR factor of b in planar

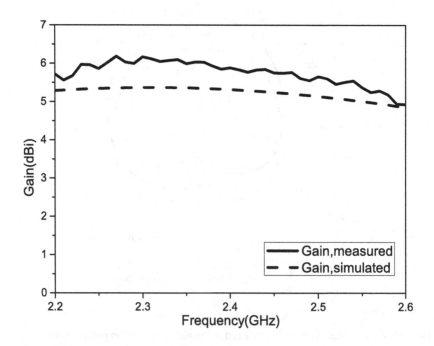

FIGURE 4.31 Simulated and measured bore-sight radiation gains of the planar self-balanced magnetic dipole antenna with a modified parasitic element.

self-balanced magnetic dipole antennas remains a critical but interesting topic. If more degree of freedom in design can be introduced, the UCR factor of b can be more flexibly tuned as desired and it may yield more novel antenna designs.

4.5 CONCLUDING REMARKS

The multi-mode resonant design approach to complementary dipole antennas has been presented and discussed in this chapter, with typical design examples raised. As is concluded, the multi-mode resonant concept can lead to high-performance complementary dipole antenna designs with the simplest configurations. Then, a novel generalized, multi-mode resonant complementary dipole antenna is advanced: The unbalanced electric current flowing on the surface of magnetic dipole antennas can be properly transformed and employed to yield planar self-balanced magnetic dipole antennas under both electric and magnetic current resonance. The key parameters of the planar self-balanced magnetic dipole can be approximately estimated by the presented closed-form design formulas (Shen et al. 2019, Wu et al. 2020). Techniques for FBR enhancement are also developed by properly adjusting the shapes of the magnetic dipole and the parasitic stub. More recently, the design approach to a planar self-balanced magnetic dipole has been successfully extended to design dual-band stacked patch antennas: As is reported, the resultant dual-band, triple-mode resonant antenna can exhibit nearly equal, widened E-plane radiation pattern with gain discrepancy less than 0.4 dB at both bands (Li et al. 2021). It is expected that the revealed operation principles and advanced design approach can be further applied in future planar, multi-mode resonant complementary dipole antenna developments.

CHAPTER 5

Multi-Mode Resonant Microstrip Patch Antennas

5.1 BRIEF HISTORY AND RECENT DEVELOPMENTS OF MICROSTRIP PATCH ANTENNAS

The concept of microstrip patch antennas (MPAs) was advanced in the early 1950s (Deschamps 1953): In that widely recognized work, a printed, light-weight microstrip feeding network was presented to feed horn antennas for the first time. Since 1955, photo-etched, strip-line configurations have been employed to design co-linear dipole array (Fubini 1955) and slot array (Sommers 1955) antennas. Later on, analyses on radiations from printed stripline discontinuities (Lewin 1960) and regularly shaped microstrip resonators (Watkins 1969) have been extensively studied through the 1960s. In these works, how to mitigate the spurious radiation effects and enhance the quality factor of microstrip configurations was more important than how to design a real, effectively radiated "microstrip antenna".

Until 1970, the first printed, planar microstrip circular patch antenna with a pair of differential coaxial feeders was pioneered (Cheston 1970), as shown in Figure 5.1a. Possibly, this might be the first printed MPA that satisfies the commonly recognized, rigorous definition of microstrip antennas (Cheston 1970): It is a planar, 2-D microstrip resonator printed on an electrically thin dielectric substrate (typically, with height less than

DOI: 10.1201/9781003291633-5

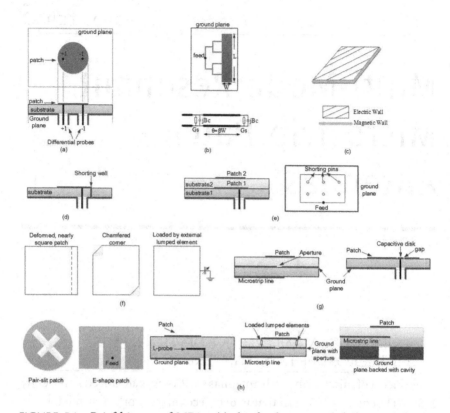

FIGURE 5.1 Brief history of MPAs: (a) the firstly reported differential-driven circular patch antenna (Cheston 1970), (b) transmission line model of MPAs (Munson 1974), (c) cavity model of MPAs (Lo et al. 1979), (d) compact patch antenna with shorting wall (Garvin et al. 1977), (e) single-mode resonant (Long and Walton 1979) and dual-mode resonant (Zhong and Lo 1983) MPAs, (f) circularly polarized (CP) MPA (Richards et al. 1981), (g) aperture-coupled (Pozar 1985) and capacitive probe compensation (Hall 1987) feeding techniques for MPAs, and (h) compact MPAs with slot-loaded (Iwasaki 1996), and wideband MPAs with lumped element loaded (Lu et al. 1998), L-probe feeding (Luk et al. 1998), E-shaped patch (Yang et al. 2001), and backed cavity (Sun and You 2010).

0.1-wavelength), which can effectively radiate by employing the equivalent magnetic currents dominated by fringe E-fields between the patch and the large ground plane. As a prevailed low-profile antenna with dozens of merits, a series of analyses and design approaches have been intensively developed since the 1970s. The most classical, milestones in the research and development of MPAs should include the transmission line model

(Munson 1974) and the microstrip cavity model (Lo et al. 1979), as given in Figure 5.1b and c: The transmission line model can offer an approximate but intuitive, equivalent circuit model for single fundamental mode resonant, rectangular MPAs analysis. The cavity model is mathematically rigorous, and it can provide clear physical insights into the operation principle, excitation, and radiation behavior of MPAs with arbitrary shapes.

A series of representative analysis approaches, feed methods, design techniques (i.e., compact size, multi-band, circularly polarized, etc.), and computer-aided design model for MPAs has been extensively developed since the late 1970s to all through the 1980s: Half-mode MPAs with short-circuited wall and compact size have been designed and used in missile mounted systems (Garvin et al. 1977), as plotted in Figure 5.1d. Stacked circular/annular patch antennas with pin-loaded and dual-band characteristics have been proposed (Long and Walton 1979, Dahele et al. 1987, Zhong and Lo 1983). As is seen from Figure 5.1e, both "multiple-radiator, single-mode resonant" (i.e., all radiators resonate at their respective principal mode) (Long and Walton 1979, Dahele et al. 1987) and "single-radiator, multi-mode resonant" (i.e., single radiator simultaneously resonates at principal and one high-order modes) (Zhong and Lo 1983) can be used to implement dual-band MPA designs. Cavity-model-based improved theory shown in Figure 5.1f has been developed for input impedance calculation, multi-port loading and excitation analysis, as well as circularly polarized (CP) MPA designs (Richards et al. 1981): The illustrated methods can be employed to perturb the magnitude and phase of a pair of degenerate, fundamental resonant modes within a square patch radiator for generating CP radiations. Eigenfunction expansion theory and generalized transmission line models have been developed to analyze the input impedance of MPAs with circular, annual ring, and other separable geometries (Chadha and Gupta 1981, Bhattacharyya and Garg 1985). Radiation behaviors of higher-order resonant modes have also been studied since then for different applications (Huang 1984, Das et al. 1984, Vaughan 1988). Besides the traditional probe- and microstrip-line-fed techniques, enhanced feeding techniques illustrated in Figure 5.1g offering more degrees of freedom in design, e.g., aperture-coupled, capacitive compensation, etc., have been developed for bandwidth enhancement (Pozar 1985, Hall 1987, Alexander 1989).

Since the 1990s, compact and miniaturization techniques for MPAs have been extensively studied with more attention, which includes some representative types:

a. Slit-loaded technique: The radius of a circular patch antenna can be reduced by about 36% by embedding a cross-slot configuration (Iwasaki 1996) into the patch's center, as seen from Figure 5.1h. A patch radiator with multiple slits may effectively miniaturize the antenna size so that it can be implemented and fabricated with microwave monolithic integrated circuits (Singh et al. 1997) on a chip. Compact, slit design is also suitable for other critically space-limited applications (Huang 2001).

b. Metallic via-hole loaded technique: Additional metallic via-holes and lumped loads (Waterhouse 1995, Lu et al. 1998, Wong 2002, Podilchak et al. 2017, Motevasselian and Whittow 2017) can also lead to minia-turized designs. Hybrid loading techniques using lumped elements can also yield compact, wideband MPA designs, while radiation efficiency may be degraded (Lu et al. 1998, Wong 2002). When a single metallic via-hole is replaced by multiple via-holes (Podilchak et al. 2017) or a short-circuited wall (Motevasselian and Whittow 2017), both planar area and height can be accordingly reduced. Parasitic radiator techniques (Ammann and Bao 2007, Wang et al. 2011, Latif et al. 2011, 2017) have been developed for miniaturization of MPAs, i.e., auxiliary radiator coupled to the principal one (Ammann and Bao 2007), resonator etched on the ground plane (Wang et al. 2011), laminated, miniaturized patch with conductor loss reduction (Latif et al. 2011, 2017), etc. Such ways can yield compact CP (Ammann and Bao 2007, Wang et al. 2011) and linearly polarized (Latif et al. 2011, 2017) antennas with enhanced performance.

c. Material-based technique: High relative permittivity (Lo et al. 1997) or permeability (Namin et al. 2010, Karilainen et al. 2011) dielectrics can be used for compact antenna size designs, while the radiation gain and efficiency may be sacrificed. Using periodic, metamaterials can possibly yield antenna miniaturization (Alù et al. 2007, Yousefi and Ramahi 2010, Yan and Vandenbosch 2016), while the resultant antenna may exhibit relatively narrowband operation.

As a planar antenna, MPA is usually recognized to have relatively low profiles much less than 0.1-wavelength: In other words, an MPA should exhibit an electrically small size in its height dimension. In addition, they often use the fundamental resonant mode for radiation only. As is well-known, a single-mode resonant MPA at the basic form exhibits a typically narrow available bandwidth of no more than 5% (Bahl and Bhartia 1980). In the past decades, the wideband technique of MPAs still "remains one of the holy grails of microstrip antenna research (Lee and Tong 2012)".

In the following section, we will introduce the novel design approaches to multi-mode resonant MPAs in the regard to "wideband" or "broadband" techniques (Kathi et al. 1987, Lee et al. 1987, Pozar and Croq 1991, Targonski et al. 1998, Ooi et al. 2002, Waterhouse 2013, Lee et al. 2017, Chiba et al. 1982, Vandenbosch and Capelle 1994, Kovitz and Rahmat-Samii 2014, Lee et al. 1997, Luk et al. 1998, Herscovici 1998, Yang et al. 2001, Sun and You 2010, Kumar and Guha 2014, Zhang et al. 2015, Mao et al. 2017, Jin et al. 2018, Bakr et al. 2019, Mcilvena and Kernweis 1979, Bernhard et al. 2003, Xiao et al. 2005, Liu et al. 2013, Liu et al. 2017, Liu et al. 2018, Liu et al. 2019), as will be seen from Figures 5.2 to 5.4.

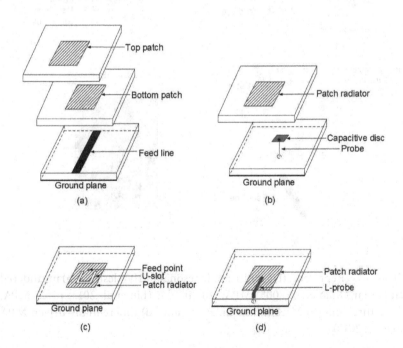

FIGURE 5.2 Representative broadband MPA designs: (a) stacked-patch antennas (Kathi et al. 1987, Lee et al. 1987, Pozar and Croq 1991), (b) patch antennas with probe capacitive compensation (Chiba et al. 1982, Vandenbosch and Capelle 1994, Kovitz and Rahmat-Samii 2014), (c) U-slot patch antennas (Lee et al. 1997), and (d) L-probe patch antennas (Luk et al. 1998).

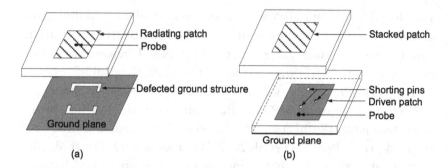

FIGURE 5.3 MPA designs incorporating external resonators and filtering functionality: (a) defected grounded structure-loaded patch antennas (Kumar and Guha 2014) and (b) filtering patch antennas (Zhang et al. 2015).

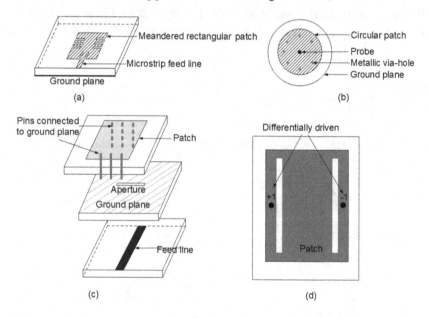

FIGURE 5.4 Representative multi-mode resonant MPA designs: (a) meandered rectangular (Xiao et al. 2005) MPA, (b) circular (Liu et al. 2013) patch MPA, (c) aperture-coupled MPA (Liu et al. 2017), and (d) differentially driven MPA (Liu et al. 2019).

A series of design approaches have been developed to enhance the available bandwidth of MPAs, with several representative examples raised in Figure 5.2. As can be seen from Figure 5.2a, a stacked-patch configuration has been proposed and widely employed in microwave- and millimeter-wave MPAs (Kathi et al. 1987, Lee et al. 1987, Pozar and Croq 1991) designs. All stacked patches should resonate at their respective fundamental mode with close frequencies at the same band, and this may lead to critically coupled wideband response with multiple resonances. As reported in the past two decades, available fractional bandwidth from 20% to 70% can be attained by using stacked patches configurations (Targonski et al. 1998, Ooi et al. 2002, Waterhouse 2013, Lee et al. 2017).

Another wideband design method for MPAs is to use thick substrate and reactance compensation techniques. In this case, the parasitic inductance of the long, coaxial feed probe may become a critical issue in wideband design considerations. Therefore, capacitive elements placed on the same layer (Chiba et al. 1982) of the patch, coupled beneath the patch (Vandenbosch and Capelle 1994), or protruded above the patch (Kovitz and Rahmat-Samii 2014) have been advanced to compensate the probe's inductance, as shown in Figure 5.2b. Alternatively, U-slot (Lee et al. 1997) (Figure 5.2c), L-probe (Luk et al. 1998) (Figure 5.2d), 3D-tapered transition (Herscovici 1998), E-shaped patch (Yang et al. 2001), and backed cavity (Sun and You 2010) configurations have been employed to compensate the inductive effect of the long probe. The capacitive compensation effect may yield a flatten reactance frequency response, such that one or multiple parasitic, less-radiative resonances can be incorporated into the principal resonant mode. In this case, only one resonant mode (i.e., the fundamental resonant mode, in general) is employed for the purpose of radiation in these designs.

Besides the stacked patches and reactance compensation techniques, external resonator loaded (Kumar and Guha 2014) and "filtenna" (filtering antenna) techniques (Zhang et al. 2015) have also been employed in wideband designs with performance enhancements (i.e., improved polarization purity, frequency selectivity in certain directions, etc.) of MPAs, as shown in Figure 5.3. The defected ground structure can be employed to enhance the impedance bandwidth and polarization purity of the MPAs (Kumar and Guha 2014). By properly incorporating shorting pins to a stacked patch antenna, less-radiative resonances can be introduced adjacent to the principal resonant mode, thus band-pass filtering functionality can be implemented without externally incorporating compensation networks (Zhang et al. 2015). More recently, a series of MPAs with various filtering functionalities have been developed by properly incorporating resonators and filtering networks to the principal radiator

(Mao et al. 2017, Jin et al. 2018, Bakr et al. 2019). As a common feature in these wideband techniques, only the fundamental resonant mode is employed for radiation. In the early MPA designs (Mcilvena and Kernweis 1979, Bernhard et al. 2003), multiple resonant modes have been utilized for radiation. Because some undesired resonant modes have not been fully suppressed, these design approaches may yield dual-band antennas instead of wideband ones.

Once some undesired resonant modes could be effectively suppressed, multiple resonant modes within a single microstrip patch radiator can be simultaneously utilized to realize a continuous, wideband radiation characteristic as expected: Orthogonal, degenerate TM_{01} and TM_{10} modes of a square patch radiator can be used for bandwidth enhancement at the cost of degraded polarization purity and gain reduction (Xiao et al. 2005), as shown in Figure 5.4a. Non-degenerate TM_{01} and TM_{02} modes within a center-fed circular patch radiator can be employed to realize a dual-mode resonant, omnidirectional radiation characteristic (Liu et al. 2013), as shown in Figure 5.4b. TM_{10}, $TM_{30,}$ and TM_{12} modes within a rectangular patch radiator can be utilized to develop multi-mode resonant, aperture-coupled, and differentially fed patch antennas (Liu et al. 2017, Liu et al. 2018, 2019), as shown in Figure 5.4c and d: Aperture-coupled feed and multiple-pin loaded technique can be employed to implement a wideband, dual-mode resonant MPA with available radiation bandwidth of 15.2% and extremely low profile of about 0.03-wavelength (Liu et al. 2017). Further improvements for cross-polarization suppression (Liu et al. 2018) and stable radiation patterns (Liu et al. 2019) can be implemented by incorporating a differentially driven feed network, to enhance the symmetry of the antenna.

In the designs shown in Figure 5.4, multiple resonant modes within a single patch radiator have been evidently excited and simultaneously employed to realize wideband designs. Unlike the reactance compensation technique based on electrically thick substrates with low relative permittivity, the inherent low-profile characteristic of MPAs can still be maintained in these multi-mode resonant designs. Therefore, the conceptual design method based upon "one radiator, multiple resonant modes" should be a promising way to effectively broaden the MPAs' bandwidth while maintaining the inherent planar, low-profile characteristic.

In the following sections of this chapter, the mode gauged design approach to MPAs based upon the generalized odd-even mode theory and multi-mode resonant concept will be systematically developed and discussed. As will be seen, the multi-mode resonant MPAs' radiation behavior can be tuned and controlled by the length of gauged magnetic dipole and the order of the resonant modes, such that novel designs with distinctive functionality can be attained.

5.2 MULTI-MODE RESONANT CIRCULAR SECTOR MPAs

5.2.1 2-D multi-Mode Resonant Magnetic Dipole Theory

Before discussing the design approach to multi-mode resonant MPAs that is based on "gauged magnetic dipole" and "mode gauged functionality", let's reveal the evolution process of 1-D linear antennas to 2-D planar antennas at first. Fundamentally, 2-D, single-mode, and multi-mode resonant circular sector MPAs can be evolved from the simplest, 1-D one-quarter wavelength, U-shaped bent dipole having quasi-isotropic radiation characteristic, as graphically illustrated in Figure 5.5. Although the stated U-shaped dipole can hardly provide sufficiently efficient radiations, it serves as an ideally analytical model in theory: As theoretically predicted and experimentally validated, a planar quasi-isotropic antenna with non-uniformity less than 5.6 dB have been successfully designed and implemented by using a planar, $TM_{2/3,1}$ mode resonant circular sector cavity (Li et al. 2017, Richards et al. 1984).

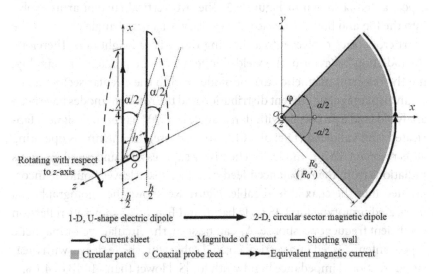

FIGURE 5.5 Evolution from 1-D U-shaped electric dipole to 2-D circular sector MPA (Li et al. 2017).

FIGURE 5.6   Photograph of a fabricated prototype of circular sector magnetic dipole antenna under $TM_{2/3,1}$ mode resonance.

A center-fed, U-shaped dipole with an approximate length of one quarter-wavelength (Liu et al. 2013) in free space shown in Figure 5.5 is considered as the original prototype. Suppose the short cross piece (with a length of h) is aligned with the z-axis, the two vertical arms can be rotated in the azimuth plane to a certain angle of α, with reference to the z-axis. In this way, a linear, 1-D electric dipole antenna may evolve into a 2-D magnetic dipole antenna shown in Figure 5.5: The two vertical current arms evolve into the top and bottom surface of a cavity with a flared angle of α, and the short cross piece evolves into a shorting wall with a height of h. Therefore, the radiation behavior of the yielded antenna can be predicted by employing the resonant magnetic current modes within the circular sector cavity. Equivalent magnetic current distribution and the resonant modes have been analyzed, and a prototype with flared angle $\alpha=270°$ has been designed, fabricated, and validated (Li et al. 2017). In this case, the antenna is operating at its resonant $TM_{2/3,1}$ mode. To effectively suppress the undesired spurious radiations from the unbalanced feed cable, a choke sleeve should be incorporated into the coaxial feed cable. Figure 5.6 shows the photograph of a fabricated prototype for 2.45 GHz band, and Figure 5.7 shows its reflection coefficient frequency response: As can be seen, the circular sector magnetic dipole antenna is fed by a vertical coaxial cable with choke sleeve, with measured fractional impedance bandwidth for $|S_{11}|$ lower than −10dB of 4.1%.

Figure 5.8 shows the measured and simulated radiation patterns in xz-, zy-, and xy-planes at 2.45 GHz in detail. In these radiation patterns,

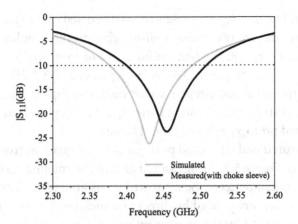

FIGURE 5.7   Simulated and measured reflection coefficients of the circular sector magnetic dipole antenna.

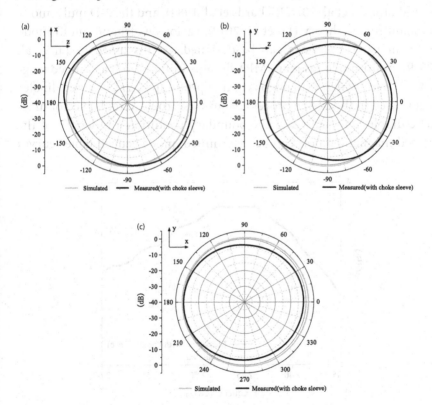

FIGURE 5.8   Simulated and measured radiation patterns of the circular magnetic dipole antenna with choke sleeve: (a) *xz*-plane, (b) *zy*-plane, and (c) *xy*-plane.

only the total field of E_{total} is presented, as discussed in (Li et al. 2017). All measured results are reasonably well-matched to the simulated ones. As depicted in Figure 5.8, the non-uniformities of measured radiation patterns are 5.1 dB in zx-plane, 5.6 dB in zy-plane, and 3.3 dB in xy-plane. Therefore, the advanced circular sector magnetic dipole antenna should exhibit a quasi-isotropic radiation pattern, just as it's poorly radiated, electric U-shaped prototype dipole should behave.

The measured and simulated peak gains of the quasi-isotropic antenna are plotted in Figure 5.9. The measured gain of the antenna varies from 1.7 to 2.9 dBi. As can be seen, the measured gains are higher than the simulated ones, which are yielded by directivity increment or non-uniformity that are contributed by the insufficiently suppressed spurious radiations, and geometrical asymmetry of the feed cable.

Based on the fractional-order resonant modes within a circular sector radiator (Li et al. 2017, Richards et al. 1984), and the 2-D multi-mode resonant dipole theory (Lu et al. 2019), various wideband, multi-mode resonant patch antennas can be designed. As discussed in Lu et al. (2019), multi-mode resonant MPA designs initiate from a "gauged magnetic dipole", which is analogous to the "prototype 1-D U-shaped dipole" that is used in the design process of quasi-isotropic, circular sector magnetic dipole antenna. Once the boundary condition and length of the gauged magnetic dipole can be determined, mode gauged tables/curves

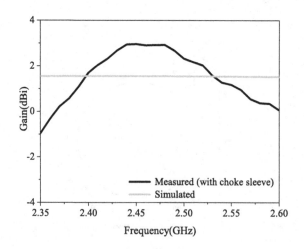

FIGURE 5.9 Simulated and measured peak gains of the circular magnetic dipole antenna with quasi-isotropic radiation patterns.

can be attained Lu et al. (2017), Lu et al. (2018). To realize wideband, multi-mode resonant MPAs, the undesired resonant modes between two usable modes should be sufficiently suppressed, or tuned out of the desired bands formed by the usable modes in preliminary. This requirement can be easily satisfied in the generalized, 2-D multi-mode resonant dipole theory, in which the undesired high-order radial modes can be suppressed by adjusting the flared angle of the circular sector radiator (Lu et al. 2018).

5.2.2 Compact Dual-Mode Resonant Circular Sector Patch Antenna

In this section, the 2-D multi-mode resonant dipole theory is employed to design a compact dual-mode resonant patch antenna. As has been reported in (Lu et al. 2018), a full-wavelength gauged magnetic dipole can be mapped into compact, dual-mode resonant circular sector patch antennas with both radii short-circuited. In that way, widened E-plane beamwidth, and available bandwidth up to 25% can be realized. Herein, we will introduce the design process of a circular sector patch antenna with a flared angle of 210° under dual-mode resonance.

As can be seen from Figure 5.10, the first step is to choose a full-wavelength gauged magnetic dipole and to map it into a circular sector radiator. The second step is to estimate the normalized radius $\bar{R}_0$, flared angle α, and the corresponding operational resonant mode. By looking up the mode gauged table in (Lu et al. 2018), it is found that when $\alpha = 210°$, the normalized radii predicted by Eqs. (5.1) and (5.2) are 0.264 and 0.273, respectively, with a discrepancy less than 0.01λ, which implies the $TM_{6/7,1}$ mode can be excited. In Eq. (5.2), $\chi_{6/7,1}$ is the first root of the first-order derivative of the 6/7-order Bessel function of the first kind. For different L, the table in (Lu et al. 2018) can be appended accordingly. A MATLAB® function for calculating the roots of $\chi_{n\pi/\alpha,\,m}$ has been presented in Appendix A for drawing the mode gauged table/chart in the symmetric gauged magnetic dipole cases (Lu et al. 2017, Lu et al. 2018).

To ensure that the second resonant mode shares identical polarization with the first usable one, the second mode should be an odd-order $TM_{18/7,1}$ mode. With reference to (Lu et al. 2017), when $TM_{\pi/\alpha,1}$ and $TM_{3\pi/\alpha,1}$ modes are simultaneously excited, the first high-order radian mode, i.e., $TM_{\pi/\alpha,2}$ mode, should be detuned to fall out of the range formed by the roots of usable modes when α is set larger than $2\pi/3$.

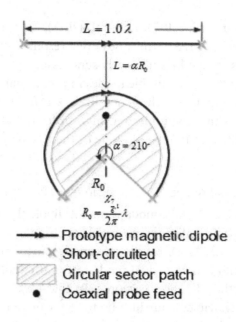

——▶▶—— Prototype magnetic dipole
———— ✕ Short-circuited
▨ Circular sector patch
● Coaxial probe feed

FIGURE 5.10 Mode mapping from a straight, full-wavelength gauged magnetic dipole to a microstrip circular sector patch radiator with both radii short-circuited to the ground plane (flared angle $\alpha = 210°$).

$$\bar{R}_0 = \frac{L}{\alpha\lambda} = \frac{1.0}{\alpha} \tag{5.1}$$

$$\bar{R}_0 = \frac{R_0}{\lambda} = \frac{\chi_{\frac{6}{7},1}}{2\pi} \tag{5.2}$$

Herein, we've attained the flared angle, radius and usable resonant modes of the circular patch antenna. The final step is to determine the feed position, and the perturbed slits' size and position. Since all usable resonant magnetic current modes exhibit cosine-dependent circumferential component, a vertical coaxial probe should be set on the x-axis ($\varphi'=0$), such that all sine-dependent, even-order modes could be fully suppressed, i.e., $\sin\nu\varphi' \equiv 0$ (Lu et al. 2018). The antenna is designed on an air substrate with height $h=5.0$ mm, at the center frequency of 2.4 GHz. To excite and tune downward the $TM_{18/7,1}$ mode, a pair of slits with a length of L_s and width W_s can be cut at $\beta=\pm35°$, where L_s and W_s can be approximately estimated as one-quarter and one-tenth wavelength of the $TM_{18/7,1}$ mode by considering, respectively. The radius of the ground plane is 62 mm, which is slightly larger than one wavelength of the center frequency (Garg et al. 2001). For better impedance matching, an additional short-circuited pin is inserted between the feed and the patch's center (Yu and Lu 2019), at the cost of patch size increment of less than 10%. Since the design guidelines have been validated to be generalized (Lu et al. 2018, Yu and Lu 2019), the prototype antenna can be designed and optimized according to the initial values. Figure 5.11 shows the photograph of a fabricated prototype.

As can be seen from Figure 5.12, the fabricated antenna exhibits a dual-mode resonant impedance bandwidth for $|S_{11}|$ lower than −10 dB from 2.35 to 3.62 GHz, which is about 45.1% in fraction. Both measured and simulated results agree with each other. Compared to the cases in (Lu et al. 2018, Yu and Lu 2019), a smaller flared angle α may lead to a larger planar area and wider impedance bandwidth. The increment of impedance bandwidth may be caused by the additional loaded shorting pin (Yu and Lu 2019).

Figure 5.13a and b illustrates the simulated surface current density distributions on the patch at 2.77 and 3.30 GHz, respectively. As can be observed, the current distribution at 2.77 GHz behaves similarly to that of the fundamental $TM_{6/7,1}$ mode. At 3.30 GHz, the surface currents tend to change direction near the slit, which implies that the $TM_{18/7,1}$ mode has been excited. Again, the dual-mode resonant principle for wideband design has been validated.

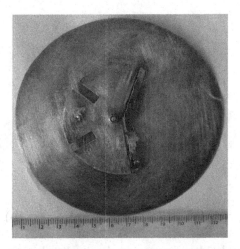

FIGURE 5.11 Photograph of dual-mode resonant wideband circular sector patch antenna with two radii short-circuited to ground plane and flared angle $\alpha = 210°$.

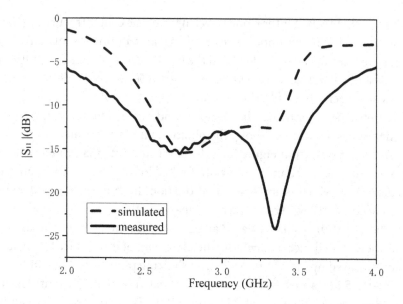

FIGURE 5.12 Simulated and measured reflection coefficients of the dual-mode resonant wideband circular sector patch antenna with flared angle $\alpha = 210°$.

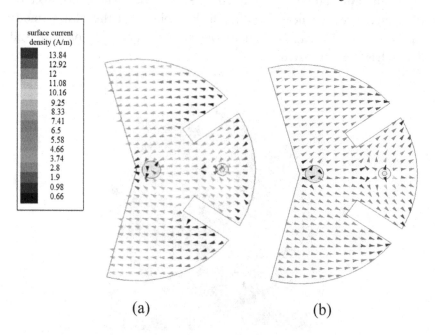

(a)   (b)

FIGURE 5.13 Simulated surface current density distributions of the dual-mode resonant wideband circular sector patch antenna with flared angle $\alpha = 210°$: (a) 2.77 GHz and (b) 3.30 GHz.

The radiation patterns at 2.77, 3.00, and 3.30 GHz are measured and compared with the simulated ones. As plotted in Figure 5.14, all measured patterns agree reasonably well with the simulated ones. Compared to the 240° patch case in (Lu et al. 2018), the prototype antenna exhibits a similar wide, E-plane (i.e., xz-plane) Half Power Beam Width (HPBW) of about 120°, and high cross-polarization level within its H-plane (i.e., yz-plane). Such a property may be beneficial for wide beamwidth antenna designs that polarization purity is regardless, e.g., radio frequency energy harvesting, etc. Compared to the single-mode resonant, quasi-isotropic design (Li et al. 2017) and dual-band, wide beam design (Mcilvena and Kernweis 1979), the dual-mode resonant patch antenna can provide a uniform, wide-angle coverage ability in the upper half zone above the ground plane within a broad impedance bandwidth up to 45.1%.

As can be seen from Figure 5.15, the measured antenna gain in the boresight, +z-direction fluctuates less than 1.4 dB within the whole impedance bandwidth. The antenna exhibits an average in-band boresight gain of 2.5 dBi, which is slightly higher than the case in (Lu et al. 2018) and comparable to that in (Yu and Lu 2019). By comparing the reflection coefficiency frequency response, radiation patterns, and gain, the antenna's available radiation bandwidth can be determined the same as its impedance bandwidth.

In this section, a wideband dual-mode resonant circular sector patch antenna has been designed by employing the mode gauged functionality (Lu et al. 2018). The radius, flared angle, feed position, perturbed slits, and usable resonant modes are regulated by a full-wavelength gauged magnetic dipole with both ends short-circuited. As can be seen, a single-layered, low-profile circular sector patch antenna with available radiation bandwidth of up to 45% can be implemented.

5.2.3 High-Gain Dual-Mode Resonant Circular Sector Patch Antenna

When both ends of a gauged magnetic dipole are set as open-circuited, the situation will be quite different. In this case, the cosine/sine dependency of the magnetic current's circumferential component will be interchanged (Lu et al. 2017, Lu et al. 2019). Alternatively, the even-order resonant modes would become cosine-dependent, and they could be excited by a coaxial probe placed at the angular bisection of $\varphi = 0$. High-gain, dual-mode resonant circular sector patch antennas can be developed by utilizing the second-order resonant mode and its fourth-order counterpart,

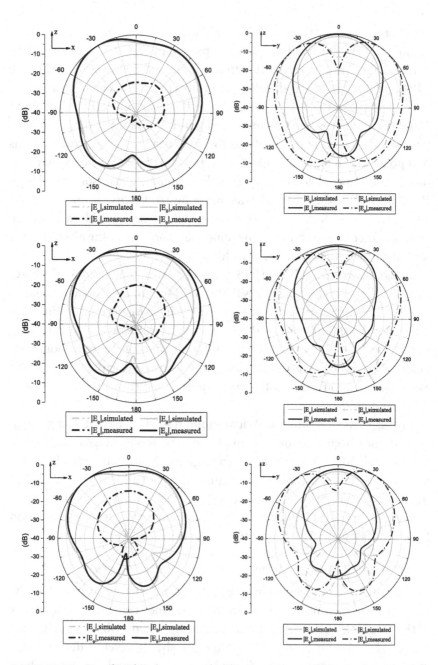

FIGURE 5.14   Simulated and measured radiation patterns of the dual-mode resonant wideband circular sector patch antenna with flared angle $\alpha = 210°$: (a) and (b) xz- and yz-planes at 2.77 GHz; (c) and (d) xz- and yz-planes at 3.00 GHz; (e) and (f) xz- and yz-planes at 3.30 GHz.

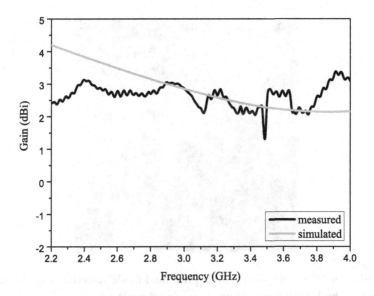

FIGURE 5.15   Simulated and measured radiation boresight gains of the dual-mode resonant wideband circular sector patch antenna with flared angle $\alpha = 210°$.

as regulated by a 1.5-wavelength gauged magnetic dipole with both radii open-circuited (Lu et al. 2017): The resultant single-layered, dual-mode resonant MPA can exhibit a boresight gain up to 10.7 dBi, low profiles from 0.02- to 0.05-wavelength, and a wide available bandwidth up to 14.5%. In this section, we will use the high-gain, dual-mode resonant circular sector patch antenna with flared angle $\alpha = 300°$ shown in Figure 5.16 to design four-element linear arrays. As illustrated in (Lu et al. 2017), the antenna should operate at its $TM_{6/5,1}$ and $TM_{12/5,1}$ modes. For size reduction, a pair of folded stubs is employed to replace the straight one in (Lu et al. 2017).

The antenna element is used to construct 4×1 linear arrays in both E- and H-planes [Shao 2020], as shown in Figure 5.17. The elements are arranged in a zig-zag configuration (Shao et al. 2019) to suppress the side lobe and back lobe, as well as to enhance the front-to-back ratio. The feed networks of both arrays are four-way, 1:1 Wilkinson power dividers. The measured and simulated reflection coefficients are plotted in Figure 5.18. As can be seen, the yielded array antennas can exhibit similar wideband, dual-mode resonant characteristics with the antenna element. The impedance bandwidth for $|S_{11}|$ lower than −10 dB for each array is larger than 10%, which satisfactorily covers the 3.4–3.6 GHz, i.e., the sub-6 GHz, a fifth-generation band in China.

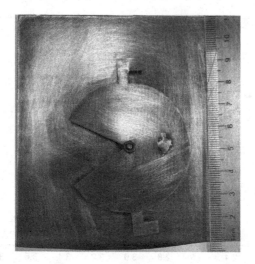

FIGURE 5.16 Photograph of a high-gain, dual-mode resonant, circular sector patch with both radii open-circuited and flared angle $\alpha = 300°$.

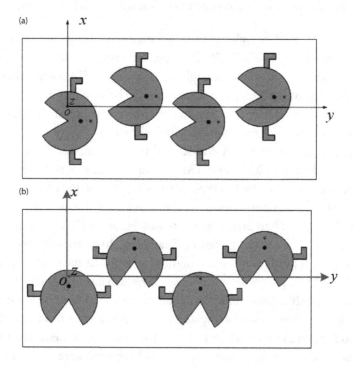

FIGURE 5.17 Dual-mode resonant, 4×1, wideband circular sector patch antenna linear arrays: (a) E-plane array and (b) H-plane array.

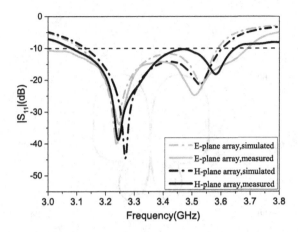

FIGURE 5.18   Simulated and measured reflection coefficients of the array antennas.

Figure 5.19 explains the reason for the choice of zig-zag array configuration by considering an E-plane array case. As can be seen, the zig-zag configuration can reduce both side and back lobe levels. As reported in (Shao et al. 2019), the measured side lobe level can be reduced to 14–15 dB lower than the main beam one, which is superior to the one of the uniform case, and the front-to-back ratio can be enhanced to 16–25 dB.

As can be seen from Figure 5.20, the boresight gain (+z-direction) of both E- and H-plane arrays is simulated, measured, and compared. Both arrays exhibit flatten, stable gain within almost the whole impedance bandwidth. The H-plane array can exhibit stable boresight gain up to 16.6 dBi, which is about 1 dB superior to the E-plane one, and more than 2 dB superior to conventional linear dipole/patch antenna arrays. At the high-frequency band from 3.5 to 3.6 GHz, the E-plane array's gain decreases to 12–13 dBi. This phenomenon may be caused by the sufficient excitation of a high order, $TM_{12/5,1}$ mode (Lu et al. 2017). The H-plane can exhibit a stable, flattened gain response within the whole impedance bandwidth. Therefore, we can conclude that the available bandwidth of the linear arrays can be determined as their impedance bandwidth.

In this section, a high-gain, dual-mode resonant circular sector patch antenna with flared angle $\alpha = 300°$ is designed and employed to construct 4×1, E- and H-plane linear arrays (Shao et al. 2019). The antenna element is laid down by a 1.5-wavelength gauged magnetic dipole with both ends open so that all even-order resonant modes can be excited by a coaxial probe. Distinctive from conventional MPAs operating at fundamental

204 ■ Multi-Mode Resonant Antennas

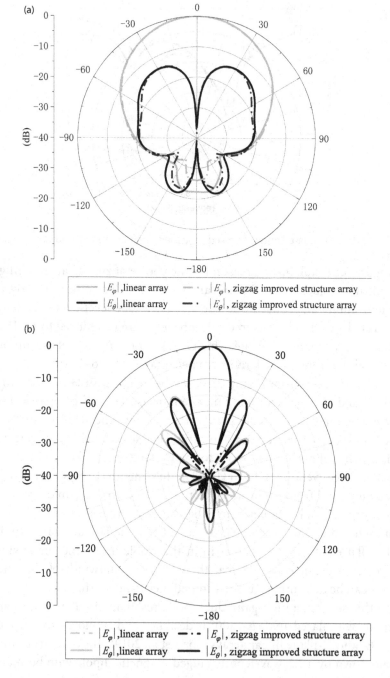

FIGURE 5.19 Simulated radiation patterns comparisons of the E-plane array at 3.25 GHz: (a) H-plane pattern and (b) E-plane pattern.

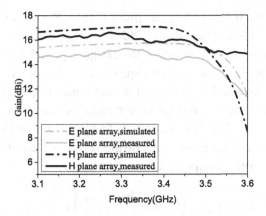

FIGURE 5.20 Simulated and measured radiation boresight gains of E- and H-plane arrays.

mode, this is an antenna operating at high-order modes only. Therefore, the antenna's equivalent aperture can be reasonably enhanced to yield a high-gain radiation pattern without exciting side and grating lobes (Lu et al. 2017). The advanced antenna has been validated to be effective in high-gain, fixed beam antenna arrays designs (Shao 2019).

5.3 MULTI-MODE RESONANT MICROSTRIP ANTENNA WITH TILTED CIRCULARLY POLARIZED BEAM AND MINIATURIZED CIRCULAR SECTOR PATCH

In the previous cases, symmetric gauged magnetic dipoles are considered. Therefore, the length of gauged magnetic dipole should be set as the integer times of one half-wavelength (Lu et al. 2019). When an asymmetric gauged magnetic dipole with one end open-circuited and the other short-circuited, CP, tilted-beam microstrip antenna under dual-mode resonance with miniaturized circular sector patch (Yu et al. 2020) can be attained. In that design, the first odd-order resonant mode and its orthogonal, second-order counterpart have been employed to generate tilted-beam circular polarization. To properly excite the desired resonant modes, the gauged magnetic dipole should be set as odd-integer multiples of one-quarter wavelength at first. Then, a microstrip circular sector configuration with one radius short-circuited and the other one open-circuited would be resulted in. As has been reported, the single-layered dual-mode resonant MPA can exhibit an impedance bandwidth of 26.6%/35.8%, low profiles from 0.046- to 0.070-wavelength, and a tilted CP 3 dB axial ratio (AR) bandwidth of 7.3%/12.0% (Yu et al. 2020).

In this section, we will introduce the design approach to tilted-beam CP patch antennas. As illustrated in Figure 5.21, 0.75- and 1.25-wavelength prototype dipole could be employed for dual, non-degenerate mode excitations within a single circular sector radiator, with one radius open-circuited and the other short-circuited. As formulated in (Yu et al. 2020), all usable resonant magnetic current modes should exhibit sine-dependent circumferential component, such that

$$M_\phi \sim M_0 J_{\frac{n\pi}{2\alpha}}(k\rho) \sin\frac{n\pi}{2\alpha}\left(\phi - \frac{\alpha}{2}\right), n = 1,3,5\ldots \tag{5.3}$$

Therefore, the feed position should be set at the open-circuited radius at $\varphi = -\alpha/2$ so that all odd-order modes can be effectively excited, as indicated by Eq. (5.3) and illustrated in Figure 5.21.

$$\begin{cases} \bar{R}_0 = \dfrac{R_0}{\lambda} = \dfrac{L}{\alpha\lambda} = \dfrac{n}{4\alpha}, n = 1,3,5\ldots \\[3mm] \bar{R}_0 = \dfrac{R_0}{\lambda} = \dfrac{\chi_{v1}}{2\pi}, v = \dfrac{n\pi}{2\alpha}, n = 1,3,5\ldots \end{cases} \tag{5.4}$$

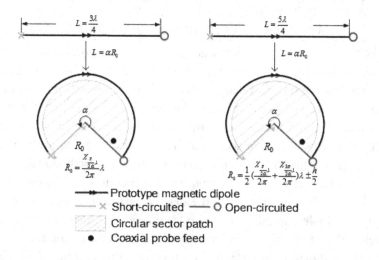

—▶▶— Prototype magnetic dipole
— ✕ Short-circuited ——O Open-circuited
░░ Circular sector patch
● Coaxial probe feed

FIGURE 5.21 Tilted-beam CP MPAs with miniaturized patch: Mode mapping from the simplest, straight, asymmetric gauged magnetic dipole to the MPAs.

$$\bar{R}_0 = \frac{\frac{\chi_{\frac{\pi}{2\alpha},1} + \chi_{\frac{3\pi}{2\alpha},1}}{}}{4\pi} \qquad (5.5)$$

The next step is to determine the usable mode, radius, position, and size of the perturbed slit. As noted in (Yu et al. 2020), when choosing a 0.75-wavelength gauged magnetic dipole, Eq. (5.4) can be employed to estimate the patch of the radius. If a 1.25-wavelength gauged magnetic dipole is employed, Eq. (5.5) may yield a more precise value since the high order mode may become more dominant in the latter case. Then, the slit position and its size can be estimated by the empirical formulas that consider the effect of the substrate's height of h (Yu et al. 2020). In final, a shorting pin (Lan and Sengupta 1985) should be incorporated near the magnetic current antinode of the high order resonant mode to provide a 90° phase shift. The separation between the pin and the center should be empirically set at about 1/3 of R_0 (Zhang and Zhu 2016c, Zhang et al. 2018). As numerically simulated, in the 240° case, the separation can be tuned to nearly one-half of R_0.

Herein, we will present a tilted-beam CP patch antenna with a flared angle of 240° to further validate the generality of the mode gauged design approach advanced in (Yu et al. 2020). The patch antenna is designed on an air substrate with height $h=5.0$ mm, at the center frequency of 2.5 GHz. A 1.25-wavelength gauged magnetic dipole has been chosen. As presented in (Yu et al. 2020), the first high-order radian mode (i.e., $TM_{\pi/2\alpha,2}$ mode) has been automatically suppressed by using a large flared angle of $\alpha > \pi/3$. A MATLAB function for calculating the roots of $\chi_{n\pi/2\alpha, m}$ has been presented in Appendix B for drawing the mode gauged table/chart in the asymmetric gauged magnetic dipole cases (Yu et al. 2020). A photograph of the fabricated prototype is shown in Figure 5.22.

The simulated and measured reflection coefficient and AR frequency responses have been plotted in Figure 5.23. As can be seen from Figure 5.23a, the impedance bandwidth for $|S_{11}|$ lower than −10dB exhibits a dual-mode resonant frequency response of 24%. Tilted-beam CP radiations can be found from 2.5 to 2.7 GHz, as indicated by the AR curves in Figure 5.23b: The ARs in $\theta=20°$ and $\theta=25°$ directions are smaller than 3 dB over 2.52 to 2.68 GHz.

The tilted-beam CP radiation is contributed by the simultaneously excited $TM_{3/8,1}$ and $TM_{9/8,1}$ modes. As illustrated by the simulated surface current distributions in Figure 5.24, the two evidently excited resonant modes

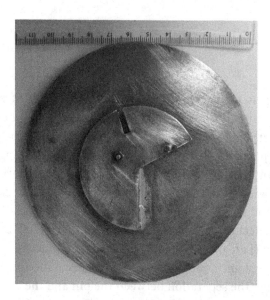

FIGURE 5.22 Photograph of a fabricated tilted-beam CP patch antenna with a flared angle of 240°.

exhibit orthogonal current distributions. Therefore, the operation principle for tilted-beam CP radiations has been numerically convinced again.

Figure 5.25 plots the left handed circularly polarized (LHCP) and right handed circularly polarized (RHCP) radiation patterns in xz-plane at three frequencies of 2.58, 2.64, and 2.70 GHz. As can be seen, the antenna exhibits an RHCP beam with a tilted angle ranging from $\theta = 17°$ to $\theta = 32°$. The antenna gain in $\theta = 20°$ direction is simulated, measured, and compared in Figure 5.26. It is seen the CP gain at 2.5–2.7 GHz band is about 2–4 dBic. It remains a challenge to enhance the CP gain in a tilted direction. To address such a critical problem, the magnitude and phase shift of E_θ and E_φ components contributed by the two usable orthogonal modes should be carefully studied. As expected, they can possibly be optimized by varying the length of gauged magnetic dipole and the flared angle. In addition, stacked patch configurations can also be employed to offer more usable modes without increasing the planar area but slightly sacrificing the antenna height to a certain extent.

In this section, miniaturized, dual-mode resonant, tilted-beam, CP microstrip circular sector patch antennas have been designed by employing the mode gauged functionality advanced in (Yu et al. 2020) and introduced in Chapter 1. As is seen, odd-integer multiples of one-quarter

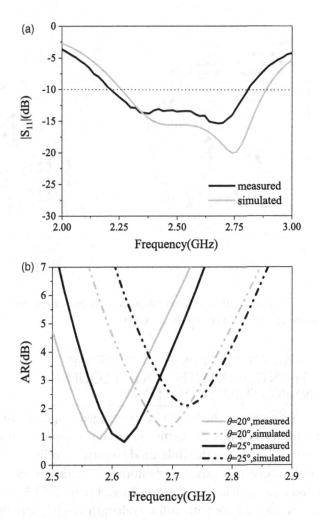

FIGURE 5.23 Simulated and measured reflection coefficients and AR of the tilted-beam CP patch antenna: (a) reflection coefficients and (b) AR.

wavelength gauged magnetic dipole can be used to determine the patch's radius, flared angle, and the usable resonant modes of a tilted-beam CP patch antenna. In addition, since the order of the eigenvalue has been reduced by one-half compared to the symmetric cases (Lu et al. 2017 and Lu et al. 2018), a miniaturized patch can also be attained.

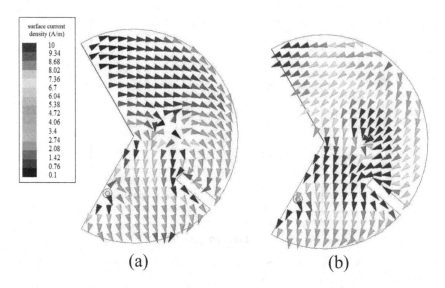

(a)   (b)

FIGURE 5.24 Simulated surface current density distributions of the tilted-beam CP patch antenna: (a) 2.4 GHz and (b) 2.7 GHz.

5.4 TRIPLE-MODE RESONANT MICROSTRIP PATCH ANTENNA WITH NULL FREQUENCY SCANNING FUNCTIONALITY

In the previous sections, the length of the gauged magnetic dipole is no more than 1.5-wavelength. As demonstrated in Chapter 3, it has been revealed that a 1.5-wavelength, triple-mode resonant slotline antenna may exhibit a null frequency scanning phenomenon within a relatively narrow elevational range (Guo et al. 2017). In addition, the 1.5-wavelength, triple-mode resonant case with full-wavelength mode partially excited also exhibits similar null frequency scanning characteristics (Wang et al. 2017). Such subtle phenomena imply that the dispersive radiation characteristic of multi-mode resonant antennas could be flexibly controlled by the length of a gauged electric or magnetic dipole, especially in a longer case. Therefore, it would be a possible way to employ a long gauged magnetic dipole to realize wideband MPAs with controllable frequency scanning functionality.

In this section, a double-wavelength gauged magnetic dipole is used to regulate and design a wideband null frequency scanning MPA. As have been reported in (Wu et al. 2020), a triple-mode resonant MPA with null frequency scanning range from elevation angle of $\theta=+41°$ to $-60°$ over

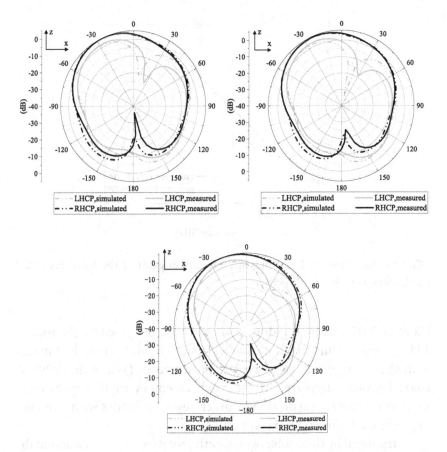

FIGURE 5.25 Simulated and measured radiation patterns of the tilted-beam CP patch antenna: (a) *xz*-plane at 2.58 GHz, (b) *xz*-plane at 2.64 GHz, and (c) *xz*-plane at 2.70 GHz.

an available radiation bandwidth of 24.2% (i.e., 2.22–2.83 GHz, for null depth smaller than −10 dB) can be designed and implemented based on the mode gauged technique of multi-mode resonant MPAs (Lu et al. 2017, 2018, Yu et al. 2020). Herein, we will study the null frequency scanning MPA with flared angle $\alpha=270°$ in further.

The antenna designs will be performed on an air substrate with relative permittivity of $\varepsilon_r=1.0$ and thickness of $h=5$ mm at 2.5 GHz band. The first step is to determine the operational modes of the circular sector MPA (Lu et al. 2018, 2017, Yu et al. 2020). The length of the gauged magnetic dipole with both ends short-circuited can be set as $L=2.0\ \lambda$, and thus lead to a circular sector MPA resonating at the $TM_{3\pi/\alpha,1}$ mode

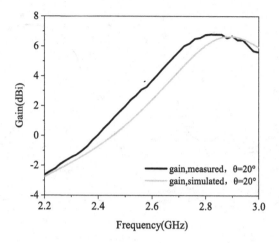

FIGURE 5.26 Simulated and measured radiation gains of the tilted-beam CP patch antenna in $\theta = 20°$ direction.

(Li et al. 2017), with its radius R_0 should be determined by the root of $TM_{3\pi/\alpha,1}$ mode, thus the first usable resonant modes can be laid down, with the aid of mode gauged chart presented in (Wu et al. 2020). A coaxial probe was placed at the patch antenna's angular bisector at $\varphi = 0$ to sufficiently excite all odd-order, resonant modes with a cosine-dependent circumferential component.

As regulated by the double-wavelength prototype dipole, it is found the third-order, $TM_{2,1}$ mode will be the principal resonant mode for radiation in the case of $\alpha = 270°$, which is distinctive to the cases in the previous sections. Next, the fundamental $TM_{2/3,1}$ mode can be perturbed up toward the principal counterpart by incorporating a pair of shorting posts (Guo et al. 2017) near the magnetic current node of the $TM_{2,1}$ mode. The position ρ_1 of the posts can be estimated by

$$J_{\frac{2}{3}}(k\rho_1) \approx J_2(k\rho_1) \qquad (5.6)$$

Then, slits can be incorporated near the outer magnetic current node of the fifth-order, $TM_{10/3,1}$ mode. The positions of the slits can be referenced to the empirical formulas by considering the substrate height h as discussed in (Wu et al. 2020). To compensate for the blind scanning zone and to implement a continuous scanning functionality, a complementary, fundamental mode resonant 90° sector radiator with a radius of R_2 that can be estimated by Eq. (5.7) would be incorporated.

$$R_2 = \frac{\chi_{2,1}}{2\pi}\lambda \qquad (5.7)$$

Therefore, a 270° circular sector patch antenna with a complementary sector radiator (flared angle of 90°) under triple-mode resonance can be designed and fabricated, as shown in Figure 5.27.

As illustrated in Figure 5.28, the measured and simulated reflection coefficients agree with each other quite well. The impedance bandwidth exhibits three resonances and ranges from 2.18–2.72 GHz. Thus a fractional impedance bandwidth for $|S_{11}|$ lower than −10 dB of 22.0% has been attained. As compared to the case presented in (Wu et al. 2020), larger flared angle α may lead to a slight reduction of about 2% in impedance bandwidth.

For further investigating the operation principle and better understanding the radiation behavior of the antenna, surface electric field distributions at 2.26, 2.47, and 2.68 GHz are simulated and plotted in Figure 5.29a–c. It is seen that the $TM_{2/3,1}$ mode has been sufficiently excited at 2.26 GHz, with the complementary radiator detuned. At 2.47 GHz, the $TM_{2,1}$ mode has become more dominant, with the complementary radiator partially excited

FIGURE 5.27   Photograph of a fabricated null frequency scanning circular sector patch antenna with complementary sector radiator.

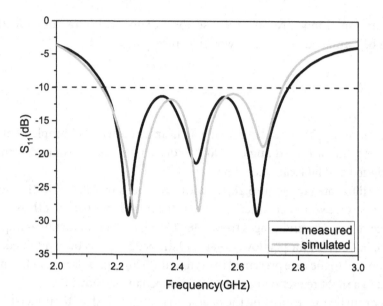

FIGURE 5.28 Simulated and measured reflection coefficients of the triple-mode resonant prototype antenna.

simultaneously. In this case, the circumferential magnetic currents flowing on the principal and complementary radiators' apertures are opposite in direction, and the separation between the apertures of two radiators is about one-wavelength centered at 2.47 GHz. Therefore, the radiations generated by the complementary radiator would be subtracted with the one by the principal radiator as the blind scanning zone has been canceled. The E-field distributions are shown in Figure 5.29 well explain the conceptual, qualitative illustration of "subtractable radiation pattern" in (Wu et al. 2020). At 2.68 GHz, the $TM_{10/3,1}$ mode has been sufficiently excited by the symmetrical slits. Till now, three usable resonant modes have been identified, illustrated to be effectively excited, and employed to attain triple-mode resonant, null frequency scanning functionality as desired.

Figure 5.30a shows and compares the simulated and measured normalized principally polarized (i.e., E_θ), E-plane (i.e., xz-plane) radiation patterns with each other. All of the normalized radiation patterns exhibit a null within xz-plane, and the null can scan from the positive elevation angle to the negative one with frequency increasing from 2.22 to 2.72 GHz (for null depth lower than −10 dB). The null can scan from $\theta=+50°$ to −64° with frequency continuously varying, without blind

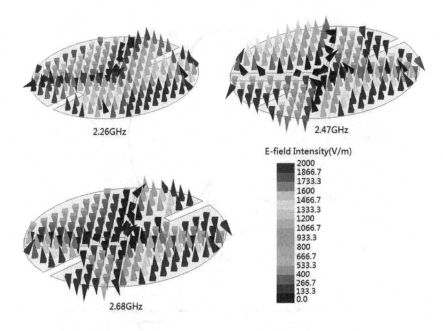

FIGURE 5.29   Simulated electric field distributions at 2.26, 2.47, and 2.68 GHz.

null scanning zone. The null frequency scanning sensitivity (NFSS) of the antenna is studied for further validating the generality of the advanced design approach (Wu et al. 2020). As can be observed from Figure 5.30b, the NFSS of the antenna exhibits an NFSS of −4.56 MHz /°. As compared to the results presented in (Wu, Yu et al. 2020), a larger flared angle may offer higher NFSS at the cost of narrower operation bandwidth. As can be seen, the 270° circular sector null frequency scanning antenna can exhibit superior null frequency performance than its 240° and 255° counterparts presented in Wu et al. (2020) and Wu, Yu et al. (2020). As comprehensive comparing with other narrowband antennas with null scanning ability (Uthansakul and Bialkowski 2006, Chatterjee et al. 2017, Yashchyshyn and Starszuk 2005, Parihar et al. 2011, Yong and Bernhard 2013, Ding et al. 2019, Dicandia et al. 2016, Labadie et al. 2012, Jiang et al. 2014, Babakhani and Sharma 2017, Iqbal and Pour 2019) in (Wu et al. 2020), the advanced triple-mode resonant antennas can exhibit a unique, continuously null frequency scanning characteristic over a wide bandwidth larger than 20%. Such distinctive null frequency scanning performance just exactly behaves as the complementary, beam scanning case of microstrip leaky-wave antennas

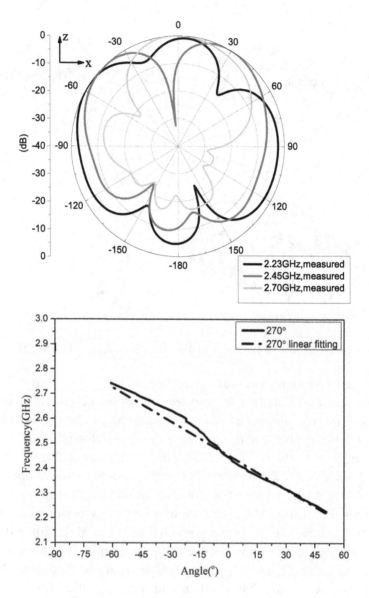

FIGURE 5.30 Null scanning characteristic of the triple-mode resonant patch antenna: (a) xz-plane radiation patterns and (b) scanning ranges/angles and NFSS.

(Menzel 1978). Unlike the counterpart that even-order resonant modes should be employed to generate a null, only odd-order resonant modes within a single patch radiator are required to realize the null frequency scanning functionality. In addition, the advanced antenna exhibits the simplest configurations that can be directly designed and implemented on a thin, single-layered air substrate, without incorporating external feed accessories for individual mode tuning. In general, the advanced design approach based upon mode gauged and double-wavelength gauged magnetic dipole can exhibit a series of merit to its counterpart. More importantly, the advanced design approach maintains the inherent low-height merit of MPAs.

In this section, a class of novel triple-mode resonant, wideband null frequency scanning circular sector patch antenna has been systematically advanced by employing the mode gauged functionality under the regulation of a double-wavelength gauged magnetic dipole. A set of closed-form design formulas has been proposed and validated. As can be seen, the proposed design approach has provided clear physical insights into planar antennas with null frequency scanning ability under triple-mode resonance (Wu et al. 2020).

5.5 CONCLUDING REMARKS

In this chapter, the brief history and recent developments of MPAs have been introduced. The 2-D multi-mode resonant magnetic dipole theory is advanced by illustrating the evolution process from 1-D, U-shaped dipole to 2-D quasi-isotropic magnetic dipole. Analogous (but not equivalent) to "prototype low-pass filter" in the synthesis process of microwave filters, a "gauged magnetic dipole" can be mapped into circular sector patch antennas and employed to regulate the order of usable modes and key parameters of the resultant antennas. In this way, the mode gauged design approach to multi-mode resonant MPAs has been advanced. It can be figured out that the behavior of the multi-mode resonant MPA should be steered by the length and boundary conditions of a gauged magnetic dipole. The order of the usable resonant modes can be finely tuned by the mode gauged functionality to the shape and key parameters of the antenna. As have been proved, the mode gauged design approach may exhibit a "mode descent" characteristic: A full-wavelength prototype dipole (Lu et al. 2018) may excite the fundamental TM mode, and the 1.5-wavelength (Lu et al. 2017) and double-wavelength (Wu et al. 2020) ones may lead to the excitation of second- and third-order TM ones, respectively. Such interesting characteristics may imply that the mode gauged approach should be beneficial for the size reduction of MPAs.

As regulated by the length and boundary condition of prototype dipole, a series of novel multi-mode resonant MPAs with distinctive characteristics has been successfully developed and investigated: Size miniaturization with wide beamwidth (Lu et al. 2018, Yu et al. 2020), high-gain (Lu et al. 2017), adjustable tilted CP beam (Yu et al. 2020), wideband null frequency scanning functionality (Wu et al. 2020, Wu, Yu et al. 2020), etc., can be accordingly attained. The advanced mode gauged design approach can provide clear physical insights into the operation principle of MPAs. Closed-form design formulas and mode gauged tables/curves (Lu et al. 2017, 2018, Yu et al. 2020) are available to precisely calculate the initial value of the key design parameters. Therefore, the mode gauged design approach would be a powerful design tool for planar antenna designers.

The mode gauged design approach yields the simplest, Pacman-shaped (Kuttler and Sigillito 1984, Bates and Ng 1973) multi-mode resonant MPAs with distinctive performance: Mathematically, the mode gauged design approach to circular sector patch antennas can be treated as the application of Fourier-Bessel series, with their eigenmodes analytically expressed by the Fourier-Bessel series in a cylindrical coordinate system. In this opinion, the mode gauged design approach can be extended in other separable curvilinear coordinate systems (Bhattacharyya and Garg 1985) or modified Pacman-shaped radiators (Bates and Ng 1973). Therefore, it is expected that the presented design approach would get wide applications in future novel MPA developments.

CHAPTER 6

Applications of Multi-Mode Resonant Antennas

6.1 APPLICATION IN ANTENNA DIVERSITY SYSTEM

As demonstrated in the previous chapters, the multi-mode resonant antenna design approach based on generalized odd-even mode theory has been systematically employed to design all kinds of resonant, elementary antennas at the simplest, basic configuration. In this chapter, more practical design examples will be raised to validate the effectiveness of the design approach and to confirm its attractive potential for all kinds of applications in wireless engineering.

The first example is the planar endfire circularly polarized antenna (PECPA) with its main beam in parallel to the antenna's plane (Lu et al. 2015; Zhang et al. 2016) under dual-mode resonance. Generally, PECPAs can be realized by using the combination of orthogonal magnetic dipoles (Lu et al. 2015) or complementary dipoles (Zhang et al. 2016). Inherently, PECPAs exhibit narrow impedance bandwidth of about 2% but a relatively wide axial ratio (AR) bandwidth higher than 10%. A dual-mode resonant straight dipole may offer an improved impedance bandwidth up to 10% (Li et al. 2016). A simple arc-shaped, dual-mode resonant dipole may enhance the impedance/AR bandwidth to 24.8%/18.8% (Zhang et al. 2016), respectively. Herein, we will use the wideband, dual-mode resonant

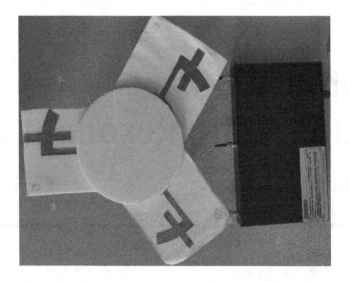

FIGURE 6.1   Co-located, dual-mode resonant, triple-PECPA diversity system for wireless router applications.

PECPA presented in Chapter 4 (Zhang et al. 2016) to build a co-located, triple-antenna diversity system for wireless routers, with prototype shown in Figure 6.1, for indoor coverage enhancement (Zhang 2017).

The port isolation of the antenna diversity system is shown in Figure 6.2a. As can be seen, although no external parasitic or filtering elements incorporated, the port isolation at the ISM-2.4 GHz band (2.404–2.484 GHz) is higher than 52 dB, which implies the antenna system can potentially provide three fully decoupled, independent circularly polarized (CP) beams and thus lead to high diversity gain. Intensive coverage range tests have been performed on seven different floors of the same building in the campus of Nanjing University of Posts and Telecommunications. In these tests, all kinds of scatterers are randomly distributed within the indoor propagation environment. The identical wireless router, personal computer, and network speed measurement software are equipped and employed to guarantee the consistency of the tests. For comparisons, the conventional case of using linearly polarized monopole antennas is also measured under the same equipment setups and in the same propagation environments. The farthest distance is defined as the distance that the personal computer could normally maintain accessing the specific network (i.e., both uploading and downloading speed are non-zero) without switching to the others. As can be seen from Figure

6.2b, the farthest available coverage distance can be enhanced by 10%–15%, compared to the conventional monopoles diversity setup (Zhang 2017). This result confirms that the dual-mode resonant PECPAs could be effectively used in enhanced-coverage indoor wireless routers.

Owning to the low-profile, wideband characteristic, multi-mode resonant PECPAs can also be potentially applied as a slim, hand-held radio-frequency identification (RFID) reader antenna (Zhang et al. 2018) or wearable antennas with spatial diversity functionality in body-area communication systems (Cui et al. 2018): Especially, a pair of dual-mode resonant PECPAs with opposite handedness have been deployed in a wearable antenna diversity system. Based on the equal-gain combination scheme and the measured propagation channel characteristics (Cui et al. 2018; Cui 2019), an off-body fading channel simulator has been developed to emulate the wearable communication performance in real, indoor fading channels under different antenna deployments. The dual-mode resonant PECPAs have been proved to be effective for fading mitigation and diversity gain enhancement in wearable communication systems: Figure 6.3a illustrates the graphical user interface of the off-body channel simulator. The path loss, time delay spread, bit error rates (BERs), and constellations of different modulation signals (BPSK, DPSK, QAM, etc.) can be simulated and displayed. Figure 6.3b shows the simulated average BER of a wearable communication system in indoor, off-body fading channels with different persons by using dual-mode resonant PECPAs, patch antennas, and planar inverted-F antennas for diversity, respectively, under the same baseband setups. As is seen, the dual-mode resonant PECPAs case exhibits the best performance with the lowest BER. This implies that dual-mode resonant PECPAs should be beneficial for body-area communication applications.

More recently, a resonant circular sector magnetic dipole has been employed to replace the rectangular leaky-wave magnetic dipole in some PECPA designs (Xue et al. 2016; Yang et al. 2018; You et al. 2016). In the future design, a multi-mode resonant circular sector magnetic dipole can be employed to simultaneously broaden both available impedance and AR bandwidth. The multi-mode resonant concept can also be further merged with widened AR beamwidth, as well as front-to-back ratio enhancement technique (You et al. 2016; Yang et al. 2018), for comprehensively improving polarization purity and available radiation bandwidth of PECPAs. More developments in this area are expected in the future.

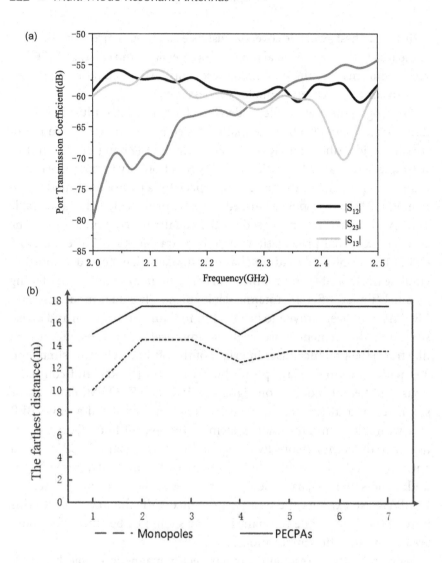

FIGURE 6.2   Performance of the diversity PECPAs system in a wireless router: (a) port isolation and (b) coverage range test results.

6.2 APPLICATION IN LOW-PROFILE ANTENNAS AND MOBILE BASE STATION ANTENNAS

In microstrip patch antenna (MPA) designs, it is always a big challenge to implement ultra-thin (i.e., no more than 0.02-wavelength) MPA designs with wide radiation bandwidth as possible. Such low-profile antennas

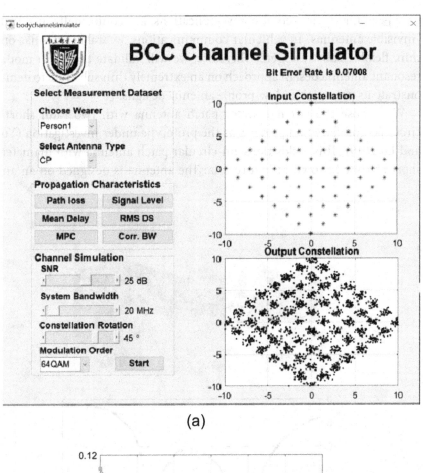

(a)

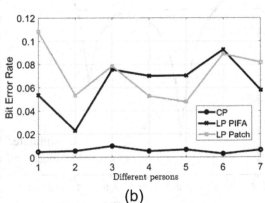

(b)

FIGURE 6.3 Wearable communication simulator using multi-mode resonant PECPAs: (a) off-body channel simulator and (b) simulated bit error rates for different diversity antennas.

have been highly desired in many applications, i.e., conformally installed "invisible antennas" in vehicular communications, wearable antennas on thin, flexible substrates, etc. Therefore, we will validate the multi-mode resonant antenna design approach on an extremely thin substrate to demonstrate its potential for low-profile antenna designs.

We choose the circular sector patch antenna with two radii short-circuited and flared angle $\alpha = \pi$ as the prototype under investigation (Yu and Lu 2019). This yields the semi-circular patch antenna with diameter short-circuited shown in Figure 6.4a: The antenna is designed on an air

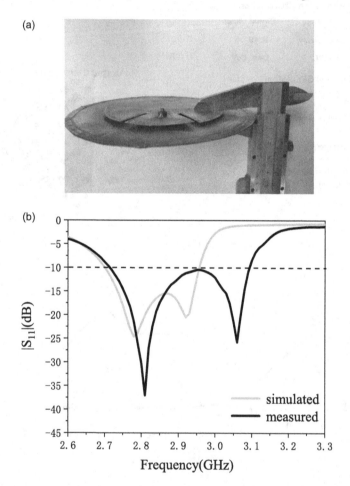

(a)

(b)

FIGURE 6.4 Dual-mode resonant, extremely thin MPA with diameter short-circuited and extremely low profile: (a) photograph of a fabricated prototype and (b) frequency responses of reflection coefficients.

substrate with a height of $h = 2$ mm, which is about 0.019-wavelength at 2.8 GHz only. In Figure 6.4a, the thickness of 1 mm of the ground plane has been included, which yields a measured height of 3 mm. Regulating by a full-wavelength gauged magnetic dipole, the principal resonant mode for radiation should be TM_{11} mode. By incorporating a pair slits near the circumferential magnetic current distribution's node, TM_{31} mode can be excited and turned downward the principal one. For better impedance matching, a short-circuited pin is loaded between the patch's center and feed point, at the cost of size increment of 9% (Yu and Lu 2019). As can be seen from Figure 6.4b, the fabricated antenna exhibits a wideband, dual-mode resonant characteristic with measured impedance bandwidth for $|S_{11}|$ lower than -10 dB from 2.72 to 3.09 GHz, which is up to 13% when the antenna height is strictly limited to 0.019-wavelength.

The xz-plane and yz-plane radiation patterns at 2.86 GHz are measured and compared with the simulated ones. As observed from Figure 6.5, the measured patterns match well with the simulated ones. In addition, the radiation patterns behave quite similar to those presented in (Yu and Lu 2019), which implies that the radiation pattern should be less frequency dispersive in the dual-mode resonant, ultra-thin profile case. As seen from Figure 6.6, the antenna exhibits a less fluctuated, dual-mode resonant bore-sight (i.e., $+z$-direction) gain frequency response from 2.72 to 3.09 GHz: The average bore-sight gain is 2.4 dBi, with fluctuation less than 0.7 dB. Therefore, the yielded ultra-thin antenna should exhibit a less frequency dispersive, wideband radiation characteristic. These results have evidently confirmed that the multi-mode resonant technique should be effective for ultra-thin, broadband MPA designs.

As is seen, the multi-mode resonant MPA on thin/ultra-thin air substrate exhibits relatively low gain (Lu et al. 2018; Yu and Lu 2019). To enhance bandwidth and gain while maintaining a relatively low-profile, stacked patch configuration can be employed (Waterhouse 1999). It is expected that a multi-mode resonant, stacked patch antenna can simultaneously provide high gain and low-profile characteristics. As shown in (Li et al. 2019), a circular patch radiator can be properly stacked to a dual-mode resonant circular sector patch antenna to yield a double-layered, triple-mode resonant MPA with a total height of about 0.086-wavelength, and impedance bandwidth of 31.5%. The in-band average bore-sight gain is up to 7.1 dBi, which is about 4.3 dB superior to that of the single-layered one (Yu and Lu 2019). However, the cross-polarization level of the stacked

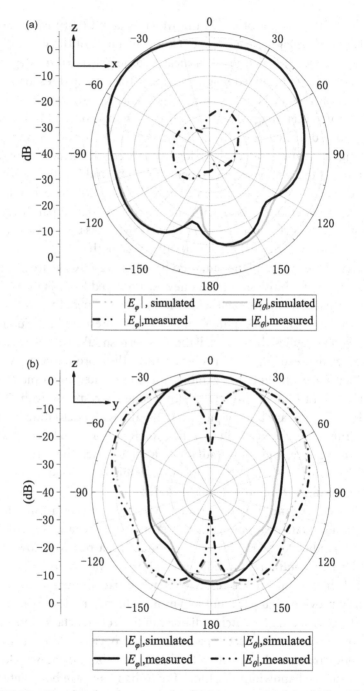

FIGURE 6.5 Simulated and measured radiation patterns of the dual-mode resonant, extremely thin MPA at 2.86 GHz: (a) *xz*-plane and (b) *yz*-plane.

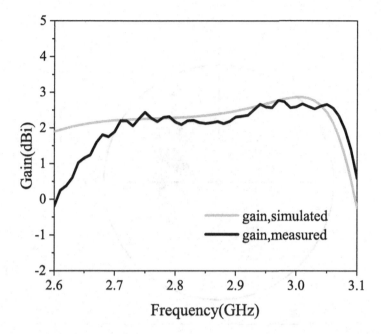

FIGURE 6.6   Simulated and measured bore-sight gain frequency response of the dual-mode resonant, extremely thin MPA.

patch antenna is relatively poor (Li et al. 2019). Besides this work, similar work using stacked dual-mode resonant, circular sector patch radiators has been advanced, as shown in Figure 6.7a. As illustrated in Figure 6.7b, dual-band, and dual-mode resonant characteristics can be attained. However, the cross-polarization level is still high, and the radiation pattern at different bands sounds quite dispersive (Li 2020). More studies and developments are expected that a multi-mode, multi-band resonant stacked patch antenna with good polarization purity would be developed in the future.

Besides the radii-short-circuited circular sector MPAs, dual-mode resonant circular sector MPAs with both radii open-circuited have been applied in high gain, fixed-beam designs with low profiles (Shao et al. 2019). The high gain characteristic is produced by the sufficiently excited, first even-order resonant mode, which can provide a larger effective aperture than that of the fundamental resonant mode. Thus the resultant antenna can exhibit higher directivity and gain, as validated in (Lu et al. 2017). Photographs of E- and H-plane 4 × 1 linear arrays using circular sector patch elements with a flared angle of 270° are presented in Figure 6.8a. All of the arrays are fabricated on an air substrate with a height of about 0.05-wavelength

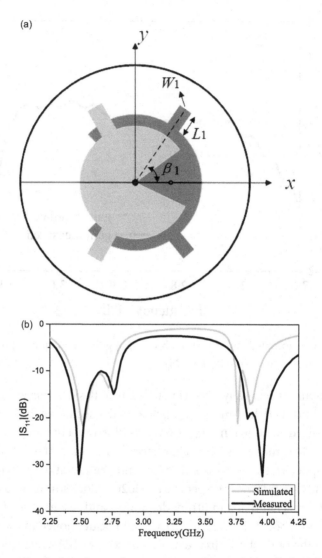

FIGURE 6.7 Stacked multi-mode resonant patch antenna with dual-band characteristic: (a) configuration and (b) impedance frequency response curves.

(Shao et al. 2019). As is seen from Figure 6.8b, both E- and H-plane arrays can exhibit bore-sight gain up to 15.5–16.3 dBi, with an enhancement of about 2 dB in comparison to a conventional array using rectangular patch elements. As presented in (Shao et al. 2019), the arrays also exhibit side-lobe levels less than −15 dB and impedance bandwidth up to 20%, which are superior to the slot counterpart (Lian et al. 2016).

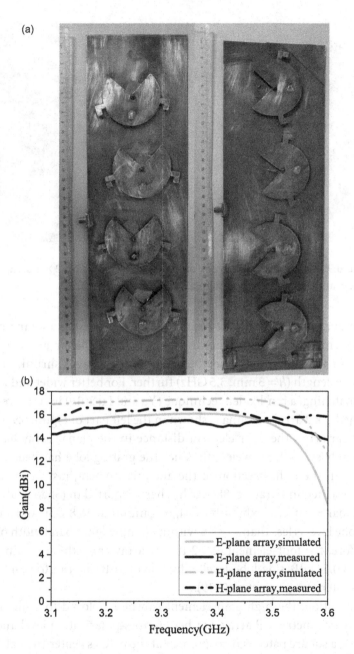

FIGURE 6.8 High-gain, fixed-beam 4 × 1 arrays using low-profile, dual-mode resonant circular sector patch antennas with flared angle of 270°: (a) photographs of fabricated E- and H-plane array prototypes and (b) simulated and measured bore-sight gain frequency responses of E- and H-plane arrays.

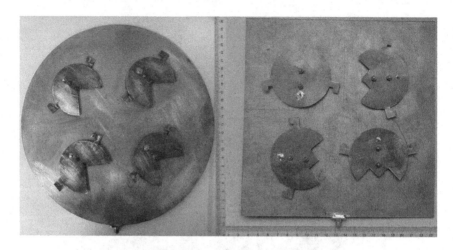

FIGURE 6.9   Fabricated prototypes of high-gain, fixed-beam 2 × 2 array antennas (left), and rotationally symmetric 2 × 2 CP array (right).

The dual even-order mode resonant, high gain patch antenna can be further employed to construct a fix-beam, 2 × 2 planar array antenna (Shao et al. 2019). In that case, the height of the array is shrunk to about 0.03-wavelength ($h = 3$ mm, 3.5 GHz) further. For better wideband impedance matching, a pin-loaded technique (Yu and Lu 2019) is employed. The designed 2 × 2 array antenna is shown in the left part of Figure 6.9. Unlike the linear array, the inter-element distance in the planar array has been reasonably set as 1.2-wavelength, since the grating lobe has been steered to exist in the null direction of the antenna element, just as the helical case presented in (Kraus 1988). As has been reported in (Shao et al. 2019), the planar array can exhibit bore-sight gain up to 16.8 dBi, and the first side-lobe level is less than −20 dB within an impedance bandwidth of 10%. Therefore, the four-element, 2 × 2 planar array can exhibit a gain increment of 0.5–1 dB, side-lobe-level reduction of 5 dB superior to its 4 × 1 linear counterpart (Shao et al. 2019).

In addition, the high gain element can be employed to build a rotationally symmetric CP array. For better cross-polarization level and gain flatness, a square patch is incorporated at the patch's center (Ji et al. 2021), as shown in the right part of Figure 6.9. Four elements are fed by a four-way, equal-magnitude, and sequential 90° phase shift Wilkinson power divider. As has been reported in (Ji et al. 2021), the yielded CP antenna can exhibit realized gain up to 15.7 (simulated) and 15.3 dBic (measured).

As is compared and seen, it can be concluded that the high gain four-element linear/planar array antennas should exhibit promising potential for serving as cost-effective base station antenna in land mobile communications. The dual, even-order mode resonant circular sector MPAs (Lu et al. 2017; Shao et al. 2019a,b et al.) should be a high gain, low-profile candidate in future wireless communications.

6.3 APPLICATION IN COMPACT MOBILE TERMINALS

In the previous sections, the multi-mode resonant concept has been applied in the designs of circularly polarized diversity antenna systems, low-profile antenna elements, and base station antenna arrays. In these designs, the environments for antenna installation are rarely considered. Wideband/multi-band antennas are highly desired in critical environments, especially in the space-limited, compact consumer electronics devices. Although the multi-mode resonant concept has been applied in all kinds of hand-held terminal antenna designs, the full-metal housings (i.e., metallic frame, and metallic back cover) of the terminal are rarely considered (Ban et al. 2015, 2016; Xu et al. 2016; et al. Elfergani et al. 2017). In this section, we will show how the multi-mode resonant concept can be used in full-metal housing mobile terminals (Li et al. 2018).

Based upon the multi-mode resonant dipole concept under asymmetric excitation (Xu et al. 2015; Lu et al. 2017), a loop-dipole antenna installed within a full-metal handset have been designed to cover most of the land mobile communication bands including GSM850/900, DCS 1800, PCS1900, UMTS2100, TD-SCDMA, LTE, etc., with its fabricated prototype shown in Figure 6.10. The fabricated prototype antenna is supported by a foam substrate with relative permittivity of nearly 1. The usable resonant modes at the basic form have been studied (Li et al. 2018). Two ring-shaped patches and a plate with stubs have been incorporated to both arms of the loop-dipole antenna, to finely tune the multi-mode resonant characteristic when it is integrated with a full metal housing, as illustrated in Figure 6.11. As can be seen, more resonances can be attained by incorporating the additional plate and stubs. The impedance bandwidths for both low- and high-frequency bands can be significantly enhanced by simultaneously exciting multiple resonant modes. For better impedance matching, a lumped, external L-shaped matching network composed of an inductor and a capacitor is introduced (Li et al. 2018). Simulated and measured results for reflection coefficients, radiation patterns, and efficiencies have been presented in

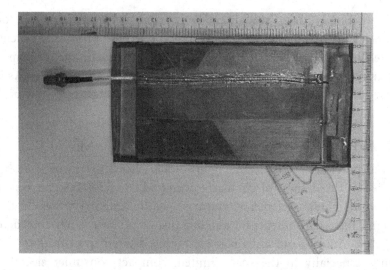

FIGURE 6.10 Photograph of a multi-mode resonant antenna in a full-metal housing mobile terminal.

(Li et al. 2018): Multi-mode resonant loop-dipole antenna packaged within a full-metal handset can exhibit usable performance for $|S_{11}|$ lower than $-6\,dB$ to cover the wireless communication spectra from 800 to 2,700 MHz. The average measured efficiencies (under passive measurement) of the

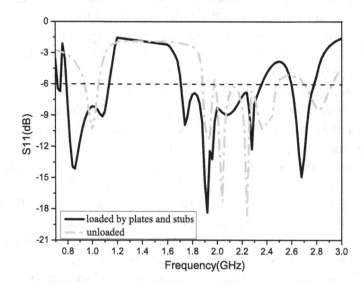

FIGURE 6.11 Comparisons of reflection coefficients under single-mode and multi-mode resonant conditions.

multi-mode resonant loop-dipole antenna are about 40%–50% and 60%–90% at 800–1,050 MHz and 1,700–2,800 MHz bands, respectively.

Under critical, zero ground clearance conditions (Gao et al. 2018), the multi-mode resonant, loop-dipole combined antenna can be integrated with the four- or eight-element, 3.5 GHz-band multiple-input-multiple-output (MIMO) handsets. The emulated fifth-generation (5G) mobile terminal prototype has been applied in real MIMO channel measurements. The case of "multi-mode resonant antenna+8-MIMO" has been shown in (Lu et al. 2019). The MIMO system is composed of slotted magnetic dipoles with single- or dual-mode resonant characteristics. The fully metallic packaged MIMO terminal has been successfully employed in the measurement of MIMO channels in the 5G mobile communication systems (She et al. 2017, 2019; Yu et al. 2016; Lu et al. 2017). As an example, emulated four-MIMO mobile terminal has been employed to measure the ergodic channel capacity in a four-Rx and eight-Tx MIMO setup scenario. As validated in (Gao et al. 2018), a four-MIMO handset can improve the capacity of the communication system with measured ergodic channel capacity up to 15.7 bps/Hz, which is 75% of the upper limit for a four-Rx, eight-Tx MIMO setup, and is nearly triple to the one of single-input-single-output case.

The aforementioned works on multi-band mobile terminal antennas have well confirmed that the multi-mode resonant concept should indeed exhibit good robustness and potential for antenna designs in extremely critical installing environments, such as fully metallic packaged mobile handsets, or other compact consume electronic devices. Recently, more developments in this area have been investigated (Xu et al. 2019; Shin et al. 2020). As illustrated in Figure 6.12a, a co-located dipole antenna under respective differential/common-mode excitation is presented. Owning to the orthogonality of differential and common modes, the co-located differential/common-mode antennas can exhibit high port isolation higher than 25 dB. It can be fabricated and integrated on the non-metal frame of the mobile terminal. This yields a good, compact four-MIMO design packaged in an emulated mobile terminal (Xu et al. 2019). A multi-branch mobile terminal antenna based upon a multi-mode resonant concept is presented in Figure 6.12b. Triple resonant modes within a planar inverted-F antenna can be simultaneously excited by incorporating a dual-branch configuration. Four or five external lumped elements are employed to compensate and match the antenna. In these ways, the available bandwidth of a planar inverted-F antenna mounted in a mobile terminal can be quadrupled or quintupled.

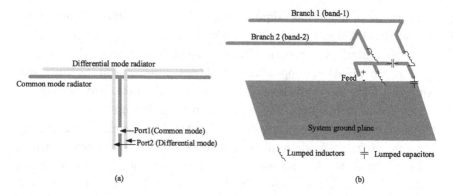

FIGURE 6.12 Recent designs of mobile terminal antennas based upon multimode resonant concept: (a) differential/common mode design (Xu et al. 2019) and (b) multi-branch design with external lumped loads (Shin et al. 2020).

6.4 APPLICATION IN IMPLANTABLE ANTENNAS

The effectiveness of the multi-mode resonant concept has been well validated in critical, fully metallic housing handsets. Another critically challenging case should be the implantable antenna designs: In this case, the antenna will be emerged within inhomogeneous human tissues with complex, non-uniform, unpredicted relative permittivity and high loss. To enhance the signal-to-noise ratio and maintain a robust intra-body wireless link, circularly polarized implantable antennas are essentially required (Liu et al. 2016, 2018; Li et al. 2017). Most of these implantable antennas are single-fed microstrip patch antennas with a pair of perturbed, degenerate mode resonance, which may suffer from their inherent narrow, single-mode resonant AR bandwidth: The narrowband, single-mode resonant AR bandwidth should be inevitably sensitive to the complex dielectrics distributed in the working environment. Thus it would be drastically changed by the complicated dielectric parameters of non-uniform tissues around the antenna. Therefore, the multi-mode resonant concept can be employed to enhance the AR bandwidth of implantable CP patch antennas (Xu et al. 2019, 2020).

As can be seen from Figure 6.13 and (Xu et al. 2019), the implantable antenna is evolved from a pin-loaded ring microstrip patch antenna. A pair of shorting pins is employed to excite the odd and even modes within the resonator. Then, slits and parasitic dipoles are diagonally incorporated to tune and perturb the non-degenerate modes. Finally, a half-mode design with a superstrate and an adjacent, emulated floating

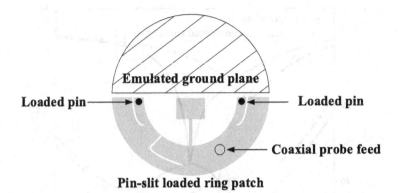

Loaded pin

Loaded pin

Emulated ground plane

Coaxial probe feed

Pin-slit loaded ring patch

FIGURE 6.13 Dual-mode resonant implantable patch antennas using pin-slit-loaded ring patch radiator with an emulated floating ground plane.

ground plane is attained (Xu et al. 2019). The impedance and AR band-widths are shown in Figure 6.14a. It is clearly seen that the antenna could exhibit dual-mode resonant AR frequency response at the ISM-2.4 GHz band. With the implanted depth varying, the AR may be degraded to about 6–7 dB; however, the dual-mode resonant frequency response is rarely changed. This confirms that the dual-resonant design approach using two non-degenerate modes should be effective for enhancing the AR bandwidth of implantable microstrip patch antenna. The AR frequency response can be further optimized by slightly modifying the slits' size in future practical developments.

To validate the antenna performances in intra-body communications, the transmitted characteristic between an external dipole and an implantable antenna is tested by using a vector network analyzer, as displayed in (Xu et al. 2019): A plastic box filled with minced pork is employed to replace the muscle phantom in numerical simulation, and the fabricated antenna is buried in the pork. Outside the muscle box in air environment, a dipole antenna with microstrip balun (operating at 2.45 GHz) is used as an external transmitter antenna to establish a wireless link with the implantable antenna. In the measurement, the dipole can be rotated by 0°, 45°, and 90° (with reference to the aligned axis of the communication link), respectively, to verify the CP characteristic of the implantable antenna. As is seen from Figure 6.15, it is seen that the rotation effect would lead to a transmitted coefficient variation of less than 3 dB when the communication distance varies from 5 to 9 cm. When the communication

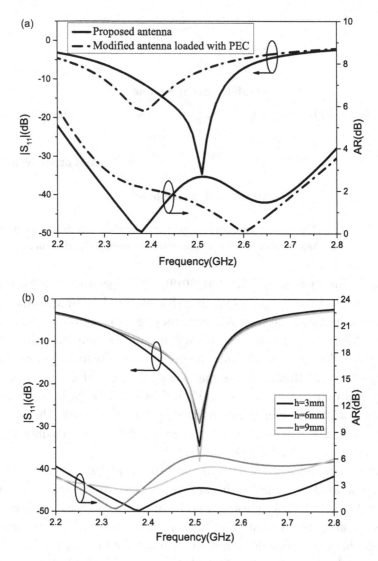

FIGURE 6.14 Impedance and AR bandwidth of the dual-mode resonant implantable antenna: (a) reflection coefficient and AR frequency response with/without an emulated ground plane and (b) effect of implanted depth on $|S_{11}|$ and AR frequency response.

distance increases to 10–20 cm, the variation may vary as high as 5–6 dB. These results indicate the implantable antenna should exhibit an AR characteristic of better than 6 dB, which would be acceptable and beneficial in practical intra-body communications.

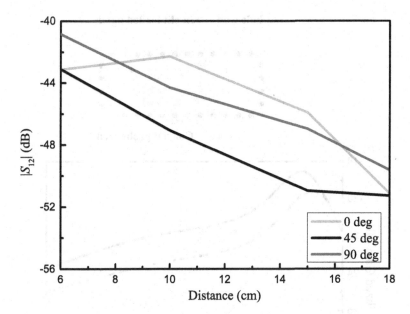

FIGURE 6.15 The effect of rotated Tx dipoles for intra-body communication link measurement.

More recently, another dual-mode resonant implantable MPA has been developed by introducing a metallic via-hole wall for both wide AR bandwidth and gain enhancement (Xu et al. 2020). In this design, the TM_{01} and TM_{02} modes within a rectangular patch radiator have been excited, perturbed, and tuned to generate a dual-mode resonant AR characteristic, as shown in Figure 6.16a. The two resonant modes are excited by two shorting pins loaded near the upper corner of the patch. An offset-fed configuration is employed to satisfy the equal-magnitude condition for CP operation. The desired 90°-phase shift is introduced by two bent slits. An additional metallic via-hole wall is surrounded to the patch for surface wave suppression and gain enhancement (Cheng et al. 2003). As is observed from Figure 6.16b, the proposed antenna can exhibit gain enhancement of 1.5 dB at the ISM-2.4 GHz band.

The proposed antenna is respectively buried in skin-mimicking agarose gel and minced pork, measured for its reflection coefficient, and compared to the simulated one, as shown in Figure 6.17. As is compared, it is seen that the antenna can exhibit dual-mode resonant characteristics in all cases.

In this section, it has been evidently confirmed that the multi-mode resonant antenna design approach can be effectively used in implantable

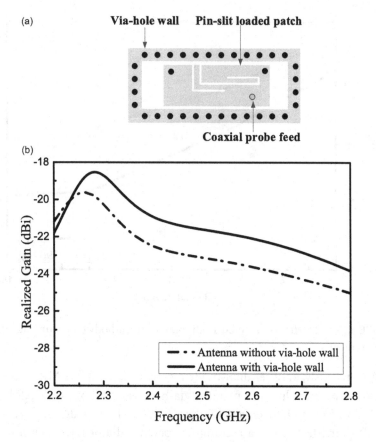

(a)   **Via-hole wall**   **Pin-slit loaded patch**

Coaxial probe feed

FIGURE 6.16 Dual-mode resonant implantable rectangular patch antenna loaded by a metallic via-hole wall: (a) antenna configuration and (b) gain enhancement comparisons.

antenna designs. This also convinces that the multi-mode resonant concept is still valid, even when the antenna is working in critically complicated environments (i.e., burying into non-uniform, dispersive, high-loss tissues with extremely complex dielectric parameters).

6.5 MISCELLANEOUS APPLICATIONS

6.5.1 Multi-Mode Resonant Dipole Antennas with Parasitic Elements for Broadcasting and Mobile Communication Applications

The multi-mode resonant concept can be further extended and merged with other antenna design approaches by properly incorporating additional, external resonators. More recently, a multi-mode resonant dipole antenna array using C-shaped parasitic element loaded dipole has been developed for digital TV transmission systems (Osklang and Phongcharoenpanich

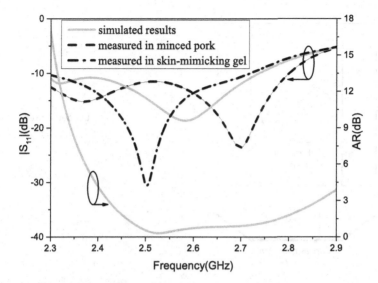

FIGURE 6.17 Simulated and measured reflection coefficient and AR frequency responses of the dual-mode resonant implantable rectangular patch antenna.

2016), as shown in Figure 6.18a. The C-shaped resonators are tightly coupled to both arms of the dipole. This yields a dual-mode resonant characteristic with fractional bandwidth up to 82.9%. The similar idea can be extended to the design of dual-polarized base station antennas, as seen from Figure 6.18b and c: Parasitic cross-dipole (Zheng and Chu 2017) and loop resonators (Wen et al. 2017) can be properly incorporated to the principal, diamond-shaped dipoles. Dual-mode resonant characteristic with available radiation bandwidth of over 52% (for voltage standing wave ratio <1.5) can be satisfactorily attained while maintaining stable beamwidth and gain variation within the whole frequency range.

More recently, a triple-mode resonant, 1.5-wavelength sectorial dipole (Pan et al. 2021) has been developed according to the multi-mode resonant, 2-D electric current sheet concept advanced in Chapter 2. It exhibits an octave available bandwidth ranging from 1.60 to 3.45 GHz, with in-band gain fluctuation less than 2.4 dB, low cross-polarization level of −27 dB, and maximum gain of 6.9 dBi. The advanced octave elementary antenna can be easily employed to construct a dual-polarized version for covering almost all-generation land mobile communication spectra in China, with one prototype shown in Figure 6.18d. Although the antenna is 1.5-wavelength in circumference, it still exhibits relative compact size owning to its unique sectorial configuration, and mode descent characteristic of its mode gauged design approach (Lu et al. 2019).

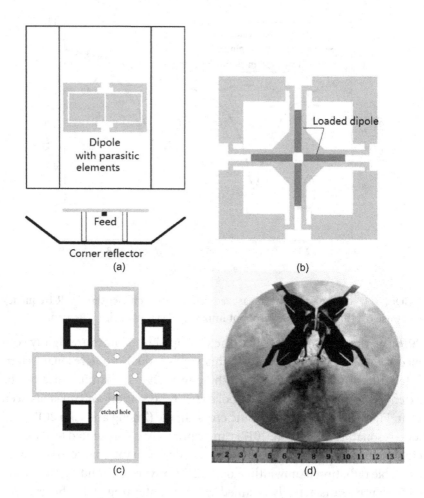

FIGURE 6.18 Multi-mode resonant antennas and arrays for TV broadcasting and mobile communications: (a) broadcasting dipole antenna array with C-shaped resonators (Osklang and Ohongcharoenpanich 2016), (b) base station antenna array with loaded cross-dipoles (Zheng and Chu 2017), and (c) loop-loaded dipole dual-polarized antenna for mobile communication base stations (Wen et al. 2017), and (d) triple-mode resonant, 1.5-wavelength, octave dipole for mobile communications (Pan et al. 2021).

6.5.2 Multi-Mode Resonant Antennas for Vehicular Communications

With the rapid development of Internet-of-Things, car-to-car (C2C) (Franck and Gil-Castineira 2007), vehicle-to-vehicle (Zhu and Roy 2003) and vehicle-to-everything (Queck et al. 2008) communications have been drawn more and more attention in recent years. As is well known to antenna designers, in all kinds of vehicular communication systems (Fujimoto and James 2000), low-profile antennas with conformal ability and low windage resistance are always welcomed and desired. As can be seen from Chapters 4 and 5, multi-mode resonant antenna design approach has been verified to be beneficial for all kinds of low-profile antennas, e.g., planar complementary dipole antennas, planar self-balanced magnetic dipole antennas, and microstrip patch antennas. Hence it would be promising to get applications in low-height, vehicular communication antenna designs.

In recent years, a multiple-mode resonant, low-profile, wideband C2C communication antenna (Wong et al. 2016) has been designed and implemented: Multiple shorting pins and V-shaped slot have been employed to simultaneously excite the TM_{10}, TM_{20}, and TM_{11} modes within an equilateral triangular patch radiator, as shown in Figure 6.19a. In this way, the antenna exhibits a triple-mode resonant, wideband characteristic and covers the bands of 5G-WLAN and C2C communications. The proposed antenna also exhibits a conical, omnidirectional beam.

Motivated by the automotive communications requirements in future intelligent transportation systems (Klemp 2010; Gallo et al. 2012), all kinds of novel, cost-effective C2C antennas and arrays with slim profile, compact size, enhanced performance and simplified configuration would be highly desired in a long period (Navarro-Mendez et al. 2017; Gao et al. 2018): As shown in Figure 6.19b, a dual-mode resonant, shorted monopole antenna with external lumped resistor and capacitor loadings have been orthogonally incorporated with a water-drop-shaped monopole antenna and integrated within a shark-fin radome, for simultaneously covering multiple spectra of the LTE700, GSM850, GSM900, DCS1800, PCS1900, WCDMA2100, WLAN2400, LTE2600,WiMAX2350, 2.4 GHz/5 GHz Wi-Fi, and, C2C bands (Navarro-Mendez et al. 2017). As can be seen from (Navarro-Mendez et al. 2017), the stubs and slits for high-order resonant mode perturbation in the multi-mode resonant antenna design approach can be replaced by lumped elements. A dual-band, center-fed annual ring/circular patch combined antenna can be designed by simultaneously exciting the TM_{01} and TM_{02} modes (Gao et al. 2018b), as shown in Figure 6.19c. The antenna can exhibit a low profile of less than 0.06-wavelength and available bandwidth of about 10% at both 2.4 GHz-WLAN and 5 GHz-WLAN/C2C bands.

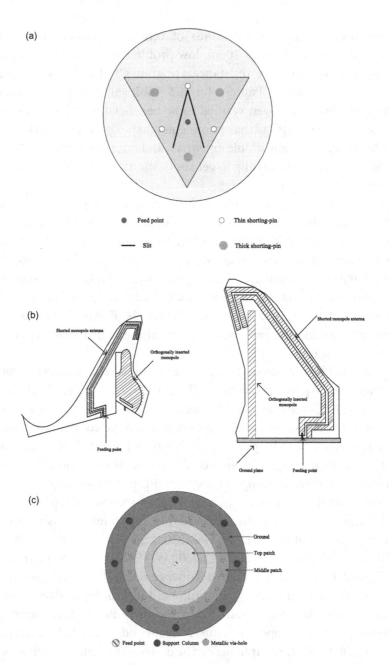

FIGURE 6.19 Multi-mode resonant antennas and arrays for vehicular communications: (a) multi-mode resonant equilateral triangular patch antenna, (b) RC-loaded, combined monopoles integrated with a shark-fin radome, and (c) concentric annual ring/circular patch antenna with multiple resonances.

6.5.3 Multi-Mode Resonant Microstrip Patch Antennas for High-Gain Applications

Another application scenario should be the high-gain, planar antennas, which can be applied in all kinds of RFID systems. The possible way for gain enhancement is to increase the antenna's effective aperture (Zhang and Zhu 2016a, 2017, 2018; Zhang et al. 2017, 2019). By incorporating shoring pins to a MPA, the resonant frequency of the principal resonant mode can be progressively tuned up so as to enlarge the electrical size of the pin-loaded patch radiator, thus enhancing its radiation directivity (Zhang and Zhu 2016a,b,c, 2017; Zhang et al. 2017) without exciting high-level side lobes. Alternatively, high-order modes can be excited, perturbed, and tuned simultaneously to form a wideband, multi-mode resonant characteristic with increased effective aperture and enhanced gain (Zhang and Zhu 2018; Zhang et al. 2019). As is shown in Figure 6.20a, four shorting pins can be properly incorporated to a square patch antenna, thus it leads to a CP MPA with a gain increment of 2.8 dB (Zhang and Zhu 2016a) to conventional patch antennas. By simultaneously exciting the high-order resonant modes within a circular or square patch, high radiation gain can be attained (Zhang and Zhu 2018; Zhang et al. 2019). In these designs, the rule of thumb is to suppress the high side-lobe level: By properly folding the out-of-phase parts of the magnetic/electric current path and suppressing their contributions to radiation, the side-lobe level can be sufficiently suppressed (Zhang and Zhu 2018). Thus high bore-sight gain up to 12–14.4 dBi can be attained (Zhang and Zhu 2018; Zhang et al. 2019). The presented multi-mode resonant, high-gain MPA design approach is expected to get application in low-profile RFID reader antenna designs.

6.5.4 Multi-Mode Resonant Liquid Antennas with Reconfigurability

The next application example is the multi-mode resonant liquid antennas with distinctive characteristics (Hua and Yang 2018; Liu et al. 2019). In general, pure water can be used to design dielectric resonator antennas (DRAs) owning to their high relative dielectric permittivity (Fayad and Record 2006). The configuration of a dual-mode resonant liquid antenna is shown in Figure 6.21a. Hybrid DRA mode and dielectric-loaded slot mode may yield a dual-mode resonant wideband characteristic in that design (Hua and Yang 2018). Thus, optical transparent, wideband antenna that is compatible to optoelectronics devices (e.g., solar cells pedal, etc.) can be implemented and integrated accordingly. A 3-D printed liquid DRA with dual-mode resonant, CP reconfigurability has been presented (Liu et al. 2019). As can be seen from Figure 6.21b, the liquid can be controlled to fill the printed container diagonally or anti-diagonally, which circularly polarized radiation characteristic with opposite handedness would be yielded. Therefore, polarization reconfigurability can be realized.

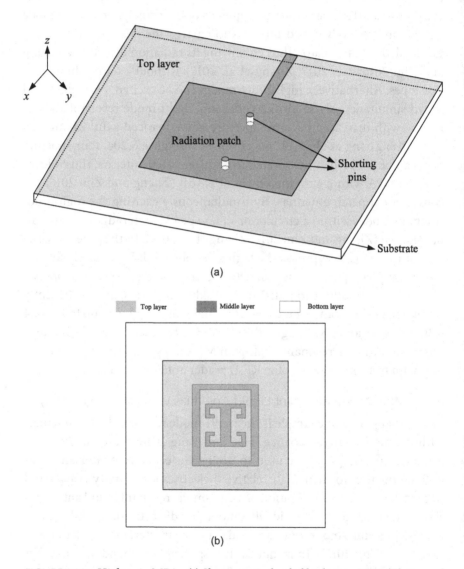

FIGURE 6.20 High-gain MPAs: (a) Shorting-pin-loaded high-gain MPA (Zhang and Zhu 2016a) and (b) high-order mode resonant high-gain MPA (Zhang et al. 2019).

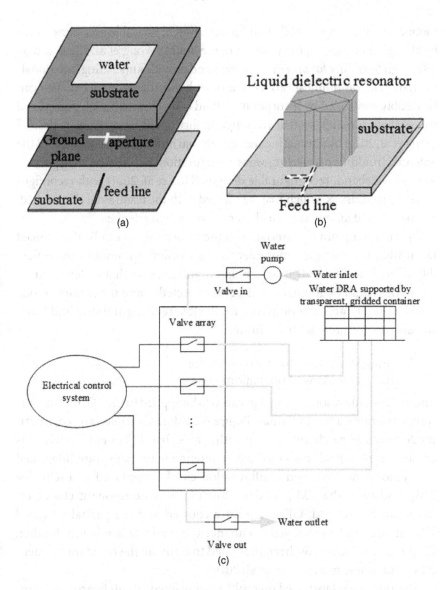

FIGURE 6.21 Multi-mode resonant liquid antennas: (a) optical transparent liquid antenna (Hua and Yang 2018), (b) 3-D printed liquid dielectric resonator antenna with reconfigurability (Liu et al. 2019), and (c) reconfigurable water DRA with a fluid control system (Xing et al. 2021).

More recently, a series of new conceptual designs have been reported in this area (Wong et al. 2020; Xing et al. 2021; Hua et al. 2021). On one hand, segment fluid antenna can be conceptually constructed on the frame of mobile terminals (i.e., mobile hand phone, watch, wearable accessories, etc.) by changing or moving the position of the liquid (Wong et al. 2020). It may offer high flexibility for antenna diversity and provide anti-fading functionality in mobile communications. On the other hand, the resonant behavior can be flexibly controlled by incorporating fluid control system with pumps and valves (Figure 6.21c), so that reconfigurability could be realized as desired (Xing et al. 2021) by alternating the operational modes and frequencies of the antenna. In addition, traveling-wave configurations can also be incorporated to yield wideband, reconfigurable designs (Hua et al. 2021) with reconfigurable characteristics, which may be merged with the traditional DRA-based, resonant liquid antennas to yield more novel design schemes.

By employing novel materials and incorporating all kinds of advanced fabrication technologies, more degrees of freedom in design can be flexibly offered to the multi-mode resonant antennas so that different functionalities could be implemented. As is expected, more novel multi-mode resonant liquid antenna designs with flexible reconfigurability and functionality would emerge in the future.

6.5.5 Multi-Mode Resonant Antennas for Millimeter-Wave Applications

The multi-mode resonant concept can also be applied in the 5G-millimeter-wave communication antennas. Figure 6.22 shows a triple-layered, multi-mode resonant patch antenna operating at K_a band (Yin et al. 2019). It is an aperture-coupled, E-shaped patch antenna fed by substrate integrated waveguide. The E-shaped configuration can be employed to excite the TM_{10} and twist the TM_{20} modes. Thus multi-mode resonant characteristics can be attained. Offset aperture coupled and the partially excited TM_{12} mode can be employed to enhance the available bandwidth further. Empirical formulas have been presented to estimate the resonant frequencies of all usable modes (Yin et al. 2019).

The antenna is fabricated on multi-layer printed circuit boards by introducing a bonding film layer to glue the E-shaped patch and the substrate integrated waveguide feed network. As is reported, the antenna can exhibit a triple-mode resonant characteristic. It covers an impedance bandwidth from 30 to 50 GHz for $|S_{11}|$ lower than −10 dB, which is 50% in fraction. The peak gain can reach 8.5 dBi, and the 3 dB gain bandwidth can reach 30%. In addition, the resultant antenna size is compact, which makes it more attractive for array designs.

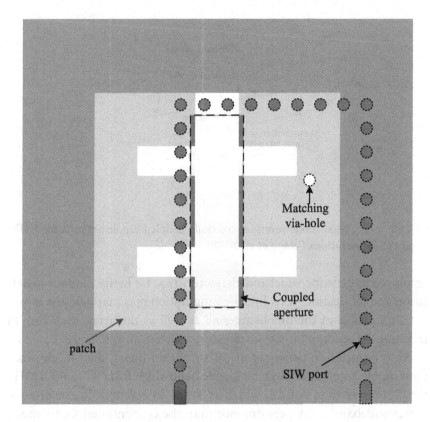

FIGURE 6.22   Multi-mode resonant E-shaped patch antenna for 5G-millimeter-wave applications (Yin et al. 2019).

Therefore, it is convinced that the multi-mode resonant concept can be applied in future millimeter-wave communication systems. Meanwhile, the multi-mode resonant concept can be incorporated with the CP millimeter-wave antennas, which may yield millimeter-wave PECPAs for mobile terminal applications with enhanced anti-fading performance (Syrytsin et al. 2018; Ruan and Chan 2020).

6.5.6 Multi-Mode Resonant Antennas for Passive Coherent Location Applications

The final example is a multi-mode resonant dipole antenna for very high frequency (VHF) band passive coherent location (PCL) applications (Wang et al. 2020). Figure 6.23 illustrates the configuration of a dual-mode resonant dipole antenna for PCL applications. Unlike the dual-mode resonant dipole discussed in Chapter 2, a coaxially coupled configuration is

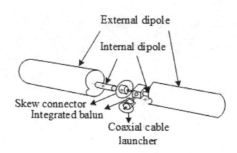

FIGURE 6.23   Dual-mode resonant coaxially coupled dipole antenna for VHF-band PCL applications (Wang et al. 2020).

employed to generate wideband characteristics, for better outdoor installation and fabrication: A co-linear configuration may maintain the inherent slim profile of the dipole antenna as well as the array, such that to facilitate the outdoor on-site installation with less air resistance.

As is reported in (Wang et al. 2020), the co-linear, dual-mode resonant dipole antenna can exhibit a wide bandwidth from 87 to 147.4 MHz for $|S_{11}|$ lower than −10 dB, which is 51.5% in fraction. It can provide better wideband array performance than the conventional single-mode resonant dipole elements. Therefore, the multi-mode resonant concept has been validated to be effective in antenna designs for low-frequency-band applications.

6.6 CONCLUDING REMARKS

In this chapter, application examples of multi-mode resonant antennas varied from VHF to millimeter-wave bands have been comprehensively raised, introduced and discussed. As is seen, the "single radiator, multiple resonant modes" concept based on the generalized odd-even mode theory has been proved to be universal, and it has been successfully applied in all kinds of antenna designs for miscellaneous applications, including mobile communications, wearable/implantable communications, vehicular communications, millimeter-wave communications, broadcasting, RFID, PCL, etc. When the antenna's height and size are critically limited, the multi-mode resonant antenna design approach still exhibits effectiveness, robustness and generality.

As is well known to most antenna designers, size reduction and bandwidth enhancement remain as big challenges in the design of planar antennas (Lee and Tong 2012). All kinds of low height, multi-mode resonant antennas face similar challenges, too. Motivated by the challenges, substantial efforts have been devoted to further enhance the performance of multi-mode resonant antennas since 2015. Lumped-distributive hybrid loading technique (Lee et al. 2015) using slits and lumped elements have been presented for size miniaturization, at the cost of radiation efficiency reduction; meandered configurations can be used for compact size design, at the cost of sacrificing antenna height and yielding high configuration complexity (Kim et al. 2017). As for a basic, rectangular multi-mode resonant microstrip patch antenna (Liu et al. 2016), to attain multiple-mode resonant characteristics, it should have to increase the size in one dimension to over one wavelength. This might possibly limit the antenna's application in array designs. For further compact design with enhanced radiation performance that can enable the antenna suitable for array applications, more comprehensive techniques by incorporating a shorting wall, slotted patch and shorting pin loadings have been further developed (Liu et al. 2017). Under a moderate low antenna height of 0.055-guided-wavelength, using V-shaped slot and double-pin loadings, the maximum occupied antenna footprint can be effectively reduced to 0.55-guided-wavelength, at the cost of narrower available bandwidth of less than 12%.

Therefore, multi-mode resonant antennas with compact size and simple configurations for all kinds of applications will always be a challenging topic for modern antenna designers (Fujimoto and Morishita 2013). It is highly expected that the multi-mode resonant antenna design approach can be merged with other advanced design and fabrication techniques (i.e., MEMS, antenna-in-package, etc.) and led to further novel compact, low-profile, integrable, and reconfigurable designs (Weedon et al. 2001; Fries et al. 2003; Gu et al. 2017). It is also expected that the presented approach can be applied in sensor antenna designs of future ubiquitous wireless networks and Internet-of-Things applications (Lu et al. 2019).

CHAPTER 7

Summarization

7.1 COMPREHENSIVE COMPARISONS TO OTHER WIDEBAND ANTENNA DESIGN APPROACHES

In the previous chapters, the wideband antenna design approaches based on generalized odd-even mode theory and "one radiator, multiple resonant modes" concept have been comprehensively introduced in detail, with elementary antenna designs and miscellaneous examples for applications. It is seen that the usable resonant modes can be clearly identified, and multi-mode resonant elementary antennas can be properly designed in a step-by-step manner: The antenna's configuration and feed position can be deduced, with all key parameters estimated and determined according to the usable resonant modes' properties. Especially, for multi-mode resonant microstrip patch antennas (MPAs), mode gauged functionality based on gauged magnetic dipole having different lengths can be advanced and employed to realize novel MPAs with distinctive performance. The gauged magnetic dipole in the mode gauged design approach to MPAs just functions analogously (but not equivalently) to the prototype low-pass filter in the synthesis method of microwave filters.

In this chapter, the multi-mode resonant antenna design approach will be systematically compared to other popular wideband antenna design approaches based on different principles or concepts. Since the 2000s, the most popular wideband antenna design approaches include the following: (1) traveling-wave, frequency-independent antenna design approach (Schelkunoff and Friis 1952; Mushiake 1996), (2) frequency- or band-notched antennas design approach (Schantz et al. 2003; Simpson 2006; Siddiqui

DOI: 10.1201/9781003291633-7

et al. 2015; Lui et al. 2005a,b, 2006, 2007, Emadian and Javad 2015), (3) filtering antenna design approach (Li et al. 2015; Duan et al. 2016a,b; Hu et al. 2016; Qian et al. 2018; Wu and Zhu 2018; Wu et al. 2018), and (4) meta-inspired antenna design approach (Dong and Itoh 2012; Liu et al. 2014). The multi-mode resonant antenna design approach based on generalized odd-even mode theory will be systematically compared to these antenna design approaches in terms of operation principle, brief closed-form design formulas for key parameters, external accessory, mode gauged functionality, and configuration complexity, as tabulated in Table 7.1.

The truncated frequency-independent antenna design approach has been compared with the multi-mode resonant one in Chapter 1. Since elaborated, smoothly tapered profiles are no longer required in the latter one, and multiple usable resonant modes can be identified, adjusted, and utilized accordingly, the multi-mode resonant antenna design approach should

TABLE 7.1 Comparison between Several Recently Developed, Popular Antenna Design Approaches

Design Approach	Operation Principle	Closed-form Design Formulas	External Accessories	Configuration Complexity	Mode Gauged Functionality
Truncated frequency-independent antennas	Truncated traveling-wave dipole	No	No	High/moderate	No
Frequency- or band-notched antennas	Band-stop filter cascaded with frequency-independent antennas	No	Yes or no	High/moderate	No
Filtering antennas	Resonant antennas with band-pass filtering functionality	No	Yes or no	High/moderate	No
Meta-inspired antennas	Leaky wave or resonant antennas with metasurfaces/periodic structures loaded	No	Yes	High/moderate	No
Multi-mode resonant antennas	Generalized odd-even mode theory	Yes	No	Low	Yes

exhibit lower structural complexity than the frequency-independent one. As commonly recognized, band-stop filtering functionality can be incorporated into a wideband antenna element, such as disk-like dipoles (Schantz et al. 2003; Simpson 2006; Siddiqui et al. 2015) or wide aperture antennas (Lui et al. 2005a,b, 2006, 2007b; Emadian and Javad 2015) by introducing one or multiple anti-resonant, narrowband parasitic element(s) to the principal radiator or the feed network. In this case, although the length of the incorporated anti-resonant, band-stop configuration can be approximately estimated as one quarter-wavelength or one half-wavelength (Schantz et al. 2003), the basic configuration cannot be theoretically predicted with all key parameters initially estimated. In addition, these designs require higher complexity in configuration and external accessories, and the order of usable modes can hardly be tuned and controlled like the ones in the multi-mode resonant antenna design approach.

Band-pass filtering functionality can be integrated into elementary antennas by embedding filter components within the radiator, to further tune gain frequency responses in a specific direction (the principal radiation direction, in general), or to reduce inter-element and trans-band couplings (Li et al. 2015; Duan et al. 2016; Hu et al. 2016; Qian et al. 2018) in arrays. This yields the concept of "filtering antennas" or "filtennas". In this case, band-pass filtering functionality can be realized by merging resonators within the principal radiator, or incorporating external resonators and synthesized filtering networks to it. Sometimes, additional accessories can be integrated within the principal radiator, to generate additional resonance near the principal mode for reactance or gain frequency response compensation (Duan et al. 2016). This implies that the radiation behavior is still dominated by the resonant fundamental mode of the principal radiator. In most cases, the filtering antenna can exhibit stable, fixed radiation patterns as conventional antennas behave. Distinctive from the filtering antennas (Li et al. 2015; Duan et al. 2016; Hu et al. 2016; Qian et al. 2018), multi-mode resonant antennas can be designed without needing to rely on external resonators. All resonant modes are identifiable or can be gauged on demand, so that they can be independently manipulated, readily employed to make their own contributions to wideband radiations. In some cases, e.g., the high-gain MPAs (Lu et al. 2017), null frequency scanning slotlines, and microstrip patch antennas (Wang et al. 2017c; Wu et al. 2020), the radiation behavior should be dominated by high-order modes rather than the fundamental one. Owning to the gauged magnetic dipole, the radiation dispersion characteristic of a multi-mode resonant antenna can be flexibly tuned and controlled. Therefore, the multi-mode resonant, wideband antenna design approach can yield both "fixed" and "dispersive" antennas as desired.

Enhanced filtering functionality has been further implemented by embedding less-radiative resonators into multi-mode resonant antennas (Wu et al. 2018; Wu and Zhu 2018) as expected: Filtering networks can be properly integrated into the principal radiator according to E-field or electric current distribution of the usable resonant mode. As is expected, more filtering antennas with diverse, enhanced performance would be accordingly designed and developed in the future. Moreover, we also expect the generalized odd-even mode theory-based, multi-mode resonant antenna design technique can emerge with other advanced techniques such as the meta-inspired antenna design approaches (Dong and Itoh 2012; Liu et al. 2014). In a more rigorous manner, the meta-inspired design approach should belong to the leaky-wave antenna design approach, which should be distinctive from the multi-mode resonant one in mathematical models. In the perspective of engineering applications, the multi-mode resonant design approach requires no externally loaded periodic structures, and it should exhibit low configuration complexity. Furthermore, brief closed-form design formulas can be deduced in the mode gauged process, to estimate the initial value of all key parameters and determine the antenna configuration in a step-by-step manner. In this opinion, electromagnetic band-gap materials and metasurfaces can be further loaded to the multi-mode resonant elementary antennas to incorporate additional resonances, and then may lead to more new, multi-mode resonant antenna designs with novel characteristics.

As comprehensively compared, it is seen that the multi-mode resonant antenna design approach based on generalized odd-even mode theory can indeed exhibit a series of distinctive advantages to the four counterparts. It is also highly expected that the multi-mode resonant antenna design approach can be merged with the other ones, to generate more novel design approaches in the future.

7.2 CONCLUSIONS

At first, a generalized odd-even mode theory is advanced to establish a common theoretical framework for resonant elementary antennas. Using the expansion of interior Green's function in terms of multiple eigenmodes, the unbalanced phenomenon in antenna systems can be mathematically modeled and quantitatively illustrated by the mutually, cross-coupled terms in the interior Green's function. The advanced theory can be employed to describe the coupled effect between antennas and feed networks (King 1943).

Then, the history, design approaches to multi-mode resonant antennas, and relevant application examples have been systematically investigated, raised, discussed, and summarized. As can be seen, multi-mode resonant elementary antennas at the simplest, basic form have been designed and validated. It is concluded that the generalized odd-even mode theory can provide a unified mathematical framework with clear physical insights for all kinds of resonant antennas: By simultaneously exciting multiple resonant modes within a single radiator, it would lead to a series of novel design approaches that exhibit generality and robustness to elementary antennas and arrays with diverse performance, as well as complicated antenna systems in all kinds of critical environments. In the previous chapters, the multi-mode resonant antennas have been proved to exhibit wideband characteristics and simple configurations. Owning to the clearly identifiable resonant modes, the multi-mode resonant antennas can be easily designed in a simple way, without the aid of extra external resonators, accessories, or lumped elements.

For most of the existing multi-mode resonant antennas at the simplest, basic form, approximate closed-form design formulas are available to estimate the initial value of key parameters. Especially, for the microstrip patch antennas, the mode gauged functionality can serve as a powerful tool for mode tuning, radiation behavior controlling, and performance optimization. The gauged magnetic or electric dipole in multi-mode resonant antenna design approaches (Lu et al. 2017; Wu et al. 2020; Yu et al. 2020; Zhao et al. 2021; Pan et al. 2021) analogously (but not equivalently) functions as the prototype low-pass filter in microwave filter synthesis approaches. Mathematically and physically, the multiple resonant modes within one radiator can be independently excited and tuned. Therefore, the radiation patterns of elementary antennas under multiple-mode resonance can be effectively controlled by optimizing the amplitude and phase distributions of the excited resonant modes within a single radiator in a shared-aperture manner. This will also lead to more novel, shared-aperture antenna designs with distinctive characteristics and diverse performance. Conventional single-mode resonant antenna design approaches can yield wideband antennas with fixed frequency dispersions, while the multi-mode resonant antenna design approach can be employed to realize wideband antennas with both fixed and adjustable, frequency dispersive radiation characteristics.

Owing to the above-mentioned attractive advantages, the generalized odd-even mode theory has established a common theoretical framework for all kinds of resonant antennas. As is expected, the multi-mode resonant antenna design approaches could get widespread applications in future antenna developments. In the future multi-mode resonant antenna developments, size miniaturization, and bandwidth broadening techniques (Lee and Tong 2012) are still challenging issues, which need to be further addressed by integrating and merging with other advanced materials and fabrication technologies (Lu et al. 2019).

Appendix A

Function for calculating the roots of the first derivative of Bessel functions:
Symmetric prototype dipole case

```
function h=pfreq4()
clc;
close all;
syms x;
c=input('Please enter the order of the Bessel J
function:');
y1=diff(besselj(c, x));
x=0.1:0.1:20;
y2=fcnchk(char(y1));
y=zeros(1,200);
for i=1:200
y(i)=y2(x(i));
end
figure
h=plot(x, y);
grid on;
a=input('down:');
b=input('up:');
y2=fcnchk(char(y1));
h=fzero(y2,[a, b]);
```

Appendix B

Function for calculating the roots of the first derivative of Bessel functions: Asymmetric prototype dipole case

```
%function bessel
%     The first root of nu-order function
clear clc
maxvl =73;
for k=1:maxvl-1
    alpha=k*pi/36;
    n=1;
    v=0.5*n*pi/alpha;
    h = v+1.9*v^(1/3)+1-2;
    fz=@(a)(-besselj(v+1,a)+v/a*besselj(v,a));
    a=0.3;
    z0(k)=fzero(fz,h);
end
z0
%     The second root of nu-order function
clear clc
maxvl =73;
for k=1:maxvl-1
    alpha=k*pi/36;
    m=0.5*pi/alpha;
    n=1;
    A=(m+1/2+2*n)*pi/2;
    B=4*m^2;
    C=7*B^2+82*B-9;
    D=83*B^3+2075*B^2-3039*B+3537;
    x(k)=A-(B+3)/(8*A)-C/(6*(4*A)^3)-D/((15*(4*A)^5));
end
x
```

References

Adekola S. A. (1983). On the excitation of a circular loop antenna by travelling and standing wave current distributions. *International Journal Electronics*, 54, 705–32.

Agrawall N. P., Kumar G., and Ray K. P. (1998). Wide-band planar monopole antennas. *IEEE Transactions on Antennas and Propagation*, 46, 294–95.

Alford A., and Kandoian A. G. (1940). Ultrahigh-frequency loop antennas. *AIEE Transactions Electrical Engineering*, 59, 843–48.

Alexander M. J. (1989). Capacitive matching of microstrip patch antennas. *IEE Proceedings, Part H*, 136, 172–74.

Alù A., Bilotti F., Engheta N., and Vegn L. I. (2007). Subwavelength, compact, resonant patch antennas loaded with metamaterials. *IEEE Transactions on Antennas and Propagation*, 55, 13–25.

Ammann M. J. and Bao X. L. (2007). Miniaturized annular ring loaded patch antenna. *IEEE Antennas and Propagation International Symposium, Honolulu, HI, USA*, 912–15.

An W. X., Wong H., Lau K. L., Li S. F., and Xue Q. (2012). Design of broadband dual-band dipole for base station antenna. *IEEE Transactions on Antennas and Propagation*, 60, 1592–95.

Babakhani B., and Sharma S. K. (2017). Dual null steering and limited beam peak steering using triple mode circular microstrip patch antenna. *IEEE Transactions on Antennas and Propagation*, 65, 3838–48.

Bahl I., and Bhartia P. (1980). *Microstrip Antennas*, London, UK: Artech House.

Bailey C. E. G. (1946). Slot feeders and slot arieals. *Journal of the Institution of Electrical Engineers, Part III A, Radiolocation*, 93, 615–19.

Bakr M. S., Großwindhager B., Rath M. et al. (2019). Compact broadband frequency selective microstrip antenna and its application to indoor positioning systems for wireless networks. *IET Microwaves. Antennas and Propagation*, 13, 1142–50.

Balanis C. A. (2005). *Antenna Theory Analysis and Design* (Third Edition), New York: John Wiley & Sons.

Ban Y.L., Qiang Y.F., Chen Z. (2015). A dual-loop antenna design for hepta-band WWAN/LTE metal-rimmed smartphone applications. *IEEE Transactions on Antennas and Propagation*, 63, 48–58.

Ban Y. L., Zhang L. W., Sim C. Y. D., Wang H., and Chen X. (2016). Heptaband coupled-fed antenna for metal-ring-frame WWAN/LTE smartphone applications. *International Journal of RF and Microwave Computer-Aided Engineering*, 26, 633–39.

Barrow W. L. (1936). Transmission of electromagnetic waves in hollow tubes of metal. *Proceedings of the IRE*, 24, 1298–1328.

Bates R.H.T. and Ng F. L. (1973). Point matching computation of transverse resonances. *International Journal for Numerical Methods in Engineering*, 6, 155–68.

Behdad N. and Sarabandi K. (2004). A multiresonant single-element wideband slot antenna. *IEEE Antennas and Wireless Propagation Letters*, 3, 5–8.

Behera A. R. and Harish A. R. (2012). A novel printed wideband dipole antenna. *IEEE Transactions on Antennas and Propagation*, 60, 4418–22.

Berndt W., and Gothe A. (1938). *Short Wave Antenna System*. US Patent 2, 138, 900.

Bernhard J. T., Mayes P. E., Schaubert D., and Mailloux R. J. (2003). A commemoration of Deschamps and Sichak's 'Microstrip Microwave Antennas': 50 years of development divergence and new directions. *27th Antenna Applications Symposium, Monticello, IL, USA*, 189–209.

Best S. R. (2010). *Progress in the Design and Realization of an Electrically Small Huygens Source*. Lisbon, Portugal: International Workshop Antenna Technology (iWAT), 1–3.

Beverage H. H. (1941). Antenna. US Patent 2, 247, 743.

Bhattacharyya A.K., and Garg R. (1985). Generalised transmission line model for microstrip patches. *IEE Proceedings, Part H*, 132, 93–98.

Blumlein A. (1939). *Improvements in or Relating to High Frequency Electrical Conductors or Radiators*. G. B. patent 515684.

Bod M., Hassani H. R. and Taheri M. M. S. (2012). Compact UWB printed slot antenna with extra bluetooth, GSM, and GPS bands. *IEEE Antennas and Wireless Propagation Letters*, 11, 531–34.

Booker H. G. (1946). Slot aerials and their relation to complementary wire aerials (Babinet's principle). *Journal of the Institution of Electrical Engineers - Part IIIA, Radiolocation*, 93, 620–26.

Brown G. H. (1936). A critical study of the characteristic of broadcast antennas as affected by antenna current distribution. *Proceedings of the IRE*, 24, 48–81.

Brown G. H. (1937). Directional antennas. *Proceedings of the IRE*, 25, 78–145.

Cai Z., Qi Y., Weng Z., Yu W., Li F., Fan J. (2018). DC ground compact wideband omnidirectional vertically polarised slot loop antenna for 4G longterm evolution applications. *IET Microwaves Antennas and Propagation*, 12, 1087–92.

Carrel R. L. (1958). The characteristic impedance of two infinite cones of arbitrary cross section. *IRE Transaction on Antennas Propagation*, AP-6, 197–201.

Carter P. S., Hansell C. W., and Lindenblad N. E. (1931). Development of directive transmitting antennas by RCA Communications, Inc. *Proceedings of the IRE*, 19, 1773–1842.

Carter P. S. (1952). *Antenna.* U.S. Patent, 2, 615, 134.

Carver K. R., and Mink J. W. (1981). Microstrip antenna technology. *IEEE Transactions on Antennas and Propagation*, AP-29, 2–24.

Chadha R., and Gupta K. C. (1981). Green's functions for circular sectors, annular rings, and annular sectors in planar microwave circuits. *IEEE Transactions on Microwave Theory and Techniques*, MTT-29, 68–71.

Chatterjee S., Chatterjee S., and Majumdar A. (2017). Edge element controlled null steering in beam-steered planar array. *IEEE Antennas and Wireless Propagation Letters*, 16, 2521–24.

Chen H. D. (2003). Broadband CPW-fed square slot antennas with a widened tuning stub. *IEEE Transactions on Antennas and Propagation*, 51, 1982–86.

Chen L., Zhang T., Wang C., and Shi X. (2014). Wideband circularly polarized microstrip antenna with wide beamwidth. *IEEE Antennas and Wireless Propagation Letters*, 13, 1577–80.

Chen M. N., Lu W. J., Wang L. J., Yang M., and Zhu L. (2019). Design approach to a novel planar bisensing circularly polarized antenna. *IEEE Transactions on Antennas and Propagation*, 67, 6839–46.

Chen X., Qin P., Guo Y. J., and Fu G. (2017). Low-profile and wide-beamwidth dual-polarized distributed microstrip antenna. *IEEE Access*, 5, 2272–80.

Chen Y., Lu W. J., Zhu L., and Zhu H. B. (2017). Square loop antenna under even-mode operation: modelling, validation and implementation. *International Journal of Electronics*, 104, 271–85.

Chen Y. B., Liu X. F., Jiao Y. C. and Zhang F. S. (2006). CPW-fed broadband circularly polarized square slot antenna. *Electronics Letters*, 42, 1074–75.

Cheng C. H., Li K., Tang K. F., and Matsui T. (2003). A new aperture-coupled patch antenna. *Microwave and Option Technology Letters*, 38, 422–23.

Cheng Y., Li Y. D., Lu W. J., and Zhu L. (2018). A wideband dual-mode complementary dipole antenna. *Electromagnetics*, 38, 134–43.

Cheston T., Byron E., and Laughlin G. (1970). Very wide-band phased arrays. *IEEE Antennas and Propagation Society. International Symposium, Columbus, OH, USA*, 226–32.

Chiba T., Suzuki Y., Miyano N., Miura S., and Ohmori S. (1982). A phased array antenna using microstrip patch antennas. *European Microwave Conference (EuMC), Helsinki, Finland*, 472–77.

Courant R., and Hilbert D. (1953). *Methods of Mathematical Physics*, (Chapters 1 and 2). Interscience Publisher.

Chu L. J. (1948). Physical limitations of omnidirectional antennas. *Journal of Applied Physics*, 19, 1163–75.

Clavin A. (1954). A new antenna feed having equal E- and H-plane patterns. *IRE Transactions on Antennas and Propagation*, AP-2, 113–19.

Clavin A. (1975). A multimode antenna having equal E- and H-planes. *IEEE Transactions on Antennas and Propagation*, AP-23, 735–37.

Collin R. E. (1991). *Field Theory of Guided Waves* (Second Edition), Chapter 2, NJ, USA: IEEE Press.

Cork E. C., and Pawsey J. L. (1939). *Wireless Aerial System*. US patent 2167709.

Cui P. -F. (2019). *Research on Short Range Wireless Off-Body Channel Propagation Characteristics and Sparse Channel Modeling (in Chinese)*. Ph.D's Dissertation, Nanjing University of Posts and Telecommunications, China.

Cui P.-F., Lu W.-J., Yu Y., Xue B., Zhu H.-B. (2018). Off-body spatial diversity reception using circular and linear polarization: measurement and modeling. *IEEE Communications Letters*, 22, 209–12.

Dahele J. S., Lee K. F., and Wong D. P. (1987). Dual-frequency stacked annular-ring microstrip antenna. *IEEE Transactions on Antennas and Propagation*, AP-35, 1281–85.

Das A., Das S. K., Mathur S. P. (1984). Radiation characteristics of higher-order modes in microstrip ring antenna. *IEE Proceedings, Part H*, 131, 102–06.

Deschamps G. A. (1951). Geometrical representation of the polarization of a plane electromagnetic wave. *Proceedings of the IRE*, 39, 540–544.

Deschamps G., and Sichak W. (1953). Microstrip microwave antennas. *3rd Symposium. USAF Antenna Res. Development. Prog. Monticello, IL, USA*.

Deschamps G. A., and Mast P. E. (1973). Poincaré sphere representation of partially polarized fields. *IEEE Transactions on Antennas and Propagation*, AP-21, 474–478.

Dicandia F. A., Genovesi S., and Monorchio A. (2016). Null-steering antenna design using phase-shifted characteristic modes. *IEEE Transactions on Antennas and Propagation*, 64, 2698–2706.

Ding C., Jones B., Guo Y. J., and Qin P. Y. (2017). Wideband matching of full-wavelength dipole with reflector for base station. *IEEE Transactions on Antennas and Propagation*, 65, 5571–76.

Ding C., Sun H. H., Zhu H., and Guo Y. J. (2020). Achieving wider bandwidth with full-wavelength dipoles (FWDs) for 5G base stations. *IEEE Transactions on Antennas and Propagation*, 68, 1119–27.

Ding Z. H., Jin R. H., Geng J. P., Zhu W. R., and Liang X. L. (2019). Varactor loaded pattern reconfigurable patch antenna with shorting pins. *IEEE Transactions on Antennas and Propagation*, 67, 6267–77.

Dong Y., and Itoh T. (2012). Metamaterial-based antennas. *Proceedings of the IEEE*, 100, 2271–85.

Duan W., Zhang X. Y., and Xue Q. (2016a). Dual-polarized high-selectivity filtering antennas without extra circuits. *International Conference on Microwave and Millimeter Wave Technology (ICMMT)*, 1–4.

Duan W., Zhang X. Y., Pan Y. M., Xu J. X., and Xue Q. (2016b). Dual-polarized filtering antenna with high selectivity and low cross polarization. *IEEE Transactions on Antennas and Propagation*, 64, 4188–96.

Duan Z. S., Qu S. B., Wu Y., and Zhang J. Q. (2009). Wide bandwidth and broad beamwidth microstrip patch antenna. *Electronics Letters*, 45, 249–50.

Dyson J. D. (1959). The unidirectional equiangular spiral antenna. *IRE Transaction on Antennas and Propagation*, AP-7, 329–34.

Elfergani I. T., Rodriguez J., Abdulsalam F., See C.H., and Alhameed A. (2017). Miniaturized balanced antenna with integrated balun for practical LTE applications. *Radioengineering*, 26, 444–52.

Emadian S. R., and Javad A.S. (2015). Very small dual band-notched rectangular slot antenna with enhanced impedance bandwidth. *IEEE Transactions on Antennas and Propagation*, 63, 4529–34.

Fayad H, Record P. (2006). Broadband liquid antenna. *Electronics Letters*, 42, 133–34.

Fenwick R. C. (1965). A new class of electrically small antennas. *IEEE Transactions on Antennas and Propagation*, AP-13, 379–83.

Franck L., and Gil-Castineira F. (2007). Using delay tolerant networks for Car2Car communications. *IEEE International Symposium on Industrial Electronics, Vigo, Spain*, 2573–78.

Fries M.K., Grani M., and Vahldieck R. (2003). A reconfigurable slot antenna with switchable polarization. *IEEE Microwave and Wireless Components Letters*, 13, 490–92.

Fubini E. G. (1955). Stripline radiator. *IRE Transactions on Microwave Theory and Techniques*, MTT-3, 149–56.

Fujimoto K., and James J. R. (2000). *Mobile Antenna Systems Handbook* (Second Edition), Boston/London: Artech House.

Fujimoto K., and Morishita H. (2013). *Modern Small Antennas*, (Chapters 7 and 8). Cambridge, UK: Cambridge University Press.

Gallo M., Bruni S., and Zamberlan D. (2012). Design and measurement of automotive antennas for C2C applications. *European Conference on Antennas and Propagation (EuCAP), Prague, Czech*, 1799–803.

Gao C., Li X.-Q, Lu W.-J., and Wong K. L. (2018a). Conceptual design and implementation of a four-element MIMO antenna system packaged within a metallic handset. *Microwave and Optical Technology Letters*, 60, 436–44.

Gao G. P., Hu B. and Zhang J. S. (2013). Design of a miniaturization printed circular-slot UWB antenna by the half-cutting method. *IEEE Antennas and Wireless Propagation Letters*, 12, 567–70.

Gao S., Ge L., Zhang D., and Qin W. (2018b). Low-profile dual-band stacked microstrip monopolar patch antenna for WLAN and car-to-car communications. *IEEE Access*, 6, 69575–581.

Garg R., Bhartia P., Bahl I., Ittipiboon A. (2001). *Microstrip Antenna Design Handbook*, Boston: Artech House.

Garvin C. W., Munson R. E., Ostwald L. T., and Schroeder K. D. (1977). Missile base mounted microstrip antennas. *IEEE Transactions on Antennas and Propagation*, AP-25, 604–10.

Ge L. and Luk K. M. (2013). A magneto-electric dipole antenna with low profile and simple structure. *IEEE Antennas and Wireless Propagation Letters*, 12, 140–42.

Geyi W., Jarmuszewski P., and Qi Y. (2000). The Foster reactance theorem for antennas and radiation Q. *IEEE Transactions on Antennas and Propagation*, 48, 401–08.

Gibson P. J. (1979). The Vivaldi aerial. *Proceedings of the 9th European Microwave Conference, Brighton, UK*, 101–05.

Goatley C., and Green F. D. (1956). Circularly-polarized biconical horns. *IRE Transactions on Antennas and Propagation*, AP-4, 592–96.

Gopikrishna M., Krishna D. D., Aanandan C. K. et al. (2009). Design of a microstip fed step slot antenna for UWB communication. *Microwave and Optical Technology Letters*, 51, 1126–29.

Granger J. V. N., and Bolljahn B. T. (1955). Aircraft antennas. *Proceedings of the IRE*, 43, 533–50.

Gu X., Liu D., Baks C., et al. (2017). A multilayer organic package with 64 dual-polarized antennas for 28GHz 5G communication. *IEEE MTT-S International Microwave Symposium (IMS), Honololu, HI, USA*, 1899–901.

Guéguen E., Thudor F., Chambelin P. (2005). A low cost UWB printed dipole antenna with high performance. *IEEE International Conference on Ultra-Wideband, Zurich, Switzerland*, 89–92.

Guo C. R., Lu W. J., Zhang Z. S., and Zhu L. (2017). Wideband non-traveling-wave triple-mode slotline antenna. *IET Microwaves Antennas and Propagation*, 11, 886–91.

Ha A., Chae M. H., and Kim K. (2019). Beamwidth control of an impulse radiating antenna using a liquid metal reflector. *IEEE Antennas and Wireless Propagation Letters*, 18, 571–75.

Hall P. S. (1987). Probe compensation in thick microstrip patches. *Electronics Letters*, 23, 606–07.

Hamid M. A. K., Boerner W. M., Shafai L., Towaij S. J., Alsip W. P., and Wilson G. J. (1970). Radiation characteristics of bent-wire antennas. *IEEE Transactions on Electromagnetic Compatibility*, EMC-12, 106–11.

Harrington R. F. (1961). *Time Harmonic Electromagnetic Fields*. (Chapter 3). USA, NY: McGraw Hill Book.

Harrison C. W., and King R. W. P. (1961). Folded dipoles and loops. *IRE Transactions on Antennas and Propagation*, AP-9, 171–87.

Herscovici N. (1998). A wide-band single layer patch antenna. *IEEE Transactions on Antennas and Propagation*, 46, 471–74.

Hines J. N., Rumsey V. H., and Walter C. H. (1953). Traveling-wave slot antennas. *Proceedings of the IRE*, 41, 1624–31.

Honda S., Ito M., Seki H., and Jinbo Y. (1992). A disk monopole antenna with 1:8 impedance bandwidth and omnidirectional radiation pattern. *International Symposium on Antennas and Propagation, Sapporo, Japan*, 1145–48.

Hsieh T. Z., Nyquist D. P., and Chen K. M. (1971). Loaded short-slot antenna with enhanced radiation or improved directivity. *Electronic Letters*, 7, 5–7.

Hsu H.-T. and Huang T.-J. (2012). Generic dipole-based antenna-featuring dual-band and wideband modes of operation. *IET Microwaves Antennas and Propagation*, 6, 1623–28.

Hu P. F., Pan Y. M., Zhang X. Y., and Zheng S. Y. (2016). A compact filtering dielectric resonator antenna with wide bandwidth and high gain. *IEEE Transactions on Antennas and Propagation*, 64, 3645–51.

Hu S. (1987). The balun family. *Microwave Journal*, 30, 227–29.

Hu W., Liu X., Gao S., Wen L., Luo Q., Fei P., Yin Y., and Liu Y. (2019). Compact wideband folded dipole antenna with multi-resonant modes. *IEEE Transactions on Antennas and Propagation*, 67, 6789–99.

Hua C., and Yang N. (2018). Optically transparent broadband water antenna. *International Journal of RF and Microwave Computer-Aided Engineering*, 28, e21219.

Hua C., Wang S., Hu Z., Zhu Z., Ren Z., Wu W. and Shen Z. (2021). Reconfigurable antennas based on pure water. *IEEE Open Journal of Antennas and Propagation*, 2, 623–33.

Huang J. (1984). Circularly polarized conical patterns from circular microstrip antennas. *IEEE Transactions on Antennas and Propagation*, AP-32, 991–94.

Huang J. (1983). A circularly polarized conical pattern from a circular microstrip antenna. *IEEE Antennas and Propagation Society International Symposium, Houston, TX, USA*, 51–54.

Huang J. (2001). Miniaturized UHF microstrip antenna for a Mars mission. *IEEE Antennas and Propagation Society International Symposium, Boston, MA, USA*, 4, 486–89.

Huang X. D., Cheng C. H. and Zhu L. (2012). An ultrawideband (UWB) slotline antenna under multiple-mode resonance. *IEEE Transactions on Antennas and Propagation*, 60, 385–89.

Iqbal Z., and Pour M. (2019). Amplitude control null steering in a multi-mode patch antenna. *Progress Electronic Research Letters*, 82, 107–12.

Itoh K. and Cheng D. K. (1972). A novel slots-and-monopole antenna with a steerable cardioid pattern. *IEEE Transactions on Aerospace and Electronic Systems*, 8, 130–34.

Iwasaki H. (1996). A circularly polarized small-size microstrip antenna with a cross slot. *IEEE Transactions on Antennas and Propagation*, 44, 1399–401.

Jang Y. W. (2000). Broadband cross-shaped microstrip-fed slot antenna. *Electronic Letters*, 36, 2056–57.

Jennetti A. G., and Uda H. (1967). Increased-directivity slot antennas. *Antennas and Propagation Society International Symposium, Ann Arbor, MI*, 64–70.

Ji F.-Y., Yu, J., and Lu W.-J. (2021). Conceptual design of high gain circularly polarized dual-resonant patch antenna array. *Applied Computational Electromagnetics Symposium, Chengdu, Sichuan, China*, 1–2.

Jia W.-Q., Ji F.-Y., Lu W. -J. et al (2021). Dual-resonant high-gain wideband Yagi-Uda antenna using full-wavelength sectorial dipoles. *IEEE Open Journal of Antennas and Propagation*, 2, 872–81.

Jiang X., Zhang Z., Li Y., and Feng Z. (2014). A novel null scanning antenna using even and odd modes of a shorted patch. *IEEE Transactions on Antennas and Propagation*, 62, 1903–09.

Jiang Z. H. and Werner D. H. (2015). A compact, wideband circularly polarized co-designed filtering antenna and its application for wearable devices with low SAR. *IEEE Transactions on Antennas and Propagation*, 63, 3808–18.

Jin J. Y., Liao S. W., and Xue Q. (2018). Design of filtering-radiating patch antennas with tunable radiation nulls for high selectivity. *IEEE Transactions on Antennas and Propagation*, 66, 2125–30.

Jin P. and Ziolkowski R. W. (2010). Metamaterial-inspired, electrically small Huygens sources. *IEEE Antennas and Wireless Propagation Letters*, 9, 501–05.

Johnson W. A. (1955). The notch aerial and some applications to aircraft radio installations. *IEE Proceedings, Part B, Radio and Electronic Engineering*, 102, 211–18.

Kahrizi M., Sarkar T. K., and Maricevic Z. A. (1993). Analysis of a wide radiating slot in the ground plane of a microstrip line. *IEEE Transactions on Microwave Theory and Techniques*, 41, 29–37.

Kandoian A. G. (1951). *Circularly Polarized Antenna*. US Patent 2539433.

Karilainen A. O., Ikonen P. M. T., Simovski C. R., and Tretyakov S. A. (2011). Choosing dielectric or magnetic material to optimize the bandwidth of miniaturized resonant antennas. *IEEE Transactions on Antennas and Propagation*, 59, 3991–98.

Kathi P. B., Alexopoulos N. G., and Hsia I. Y. (1987). A bandwidth enhancement method for microstrip antennas. *IEEE Transactions on Antennas and Propagation*, AP-35, 5–12.

Khraisat Y., Hmood K., and Anwar A. (2012). Analysis of the radiation resistance and gain of full-wave dipole antenna for different feeding design. *Journal of Electromagnetic Analysis and Applications*, 4, 235–42.

Kim J.-H., Jeong M.-G., Bae S.-H., and Lee W.-S. (2017). A printed fan-shaped meandered dipole antenna with mutual-coupled dual resonance. *IEEE Antennas and Wireless Propagation Letters*, 16, 3168–71.

Kimouche H., Abed D., and Atrouz B. (2010). Small size microstrip slot antennas for ultra-wideband communications. *European Conference on Antennas and Propagation (EuCAP), Barcelona, Spain*, 1–5.

King R. (1943). Coupled antennas and transmission lines. *Proceedings of the IRE*, 31, 626–40.

King R. W. P. (1956). *Theory of Linear Antennas*. Cambridge, MA: Harvard University Press.

King R. W. P. and Owyang G. H. (1960). The slot antenna with coupled dipoles. *IRE Transactions on Antennas and Propagation*, AP-8, 136–43.

King R., Fikioris G. J., and Mack R. B. (2002). *Cylindrical Antennas and Arrays*. Cambridge, UK: Cambridge University Press.

Klemp O. (2010). Performance considerations for automotive antenna equipment in vehicle-to-vehicle communications. *URSI International Symposium on Electromagnetic Theory (EMTS), Berlin, Germany*, 934–37.

Ko S., and Lee J. (2013). Hybrid zeroth-order resonance patch antenna with broad E-plane beamwidth. *IEEE Transactions on Antennas and Propagation*, 61, 19–25.

Kovitz J. M., and Rahmat-Samii Y. (2014). Using thick substrates and capacitive probe compensation to enhance the bandwidth of traditional CP patch antennas. *IEEE Transactions on Antennas and Propagation*, 62, 4970–79.

Kraus J. D. (1940). The corner reflector antenna. *Proceedings of the IRE*, 28, 513–19.

Kraus J. D. (1948). Helical beam antennas for wide-band applications. *Proceedings of the IRE*, 36, 1236–42.

Kraus J. D. (1988). *Antennas* (Second Edition), Chapter 8. Boston, MA: McGraw-Hill.

Kraus J. D., and Marhefka R. J. (2002). *Antennas: For All Applications*. New York: McGraw-Hill.

Kumar C., and Guh D. A. (2014). Defected ground structure (DGS)-integrated rectangular microstrip patch for improved polarisation purity with wide impedance bandwidth. *IET Microwaves Antennas and Propagation*, 8, 589–96.

Kuo F.Y., Chou H. T., Hsu H. T., et al. (2010). A novel dipole antenna design with an over 100% operational bandwidth. *IEEE Transactions on Antennas and Propagation*, 58, 2737–41.

Kuo Y.-L. and Wong K.-L. (2003). Printed double-T monopole antenna for 2.4/5.2 GHz dual-band WLAN operations. *IEEE Transsctions on Antennas and Propagation*, 51, 2187–92.

Kuttler J. R., and Sigillito V. G. (1984). Eigenvalues of the Laplacian in two dimensions. *SIAM Review*, 26, 163–93.

Lai A. K. Y., Sinopoli A. L., and Burnside W. D. (1992). A novel antenna for ultra-wideband applications. *IEEE Transactions on Antennas and Propagation*, 40, 755–60.

Labadie N. R., Sharma S. K., and Rebeiz G. (2012). Multimode antenna element with hemispherical beam peak and null steering. *IEEE Antennas and Propagation Society International Symposium, Chicago, IL, USA*, 1–2.

Lan G. L., and Sengupta D. (1985). Post-tuned single-feed circularly polarized patch antennas. *IEEE Antennas & Propagation Society International Symposium, Vancouver, BC, Canada*, 85–88.

Landsdorfer F. M. (1976). Zur optimalen form von linearantennen (in German). *Frequenz*, 30, 344–49.

Langley J. D. S., Hall P. S., and Newham P. (1993). Novel ultra-wideband Vivaldi antenna with low crosspolarisation. *Electronics Letters*, 29, 2004–05.

Latif S. I., Shafai L. and Sharma S. K. (2005). Bandwidth enhancement and size reduction of microstrip slot antennas. *IEEE Transactions on Antennas and Propagation*, 53, 994–1003.

Latif S.I., Shafai L., and Shafai C. (2011). Gain and efficiency enhancement of compact and miniaturised microstrip antennas using multi-layered laminated conductors. *IET Microwaves Antennas and Propagation*, 5, 402–11.

Latif S.I., Shafai L., and Shafai C. (2017). An engineered conductor for gain and efficiency improvement of miniaturized microstrip antennas. *IEEE Antennas and Propagation Magazine*, 55, 77–90.

Le D. T. and Karasawa Y. (2012). A novel compact ultra-wideband dipole antenna. *European Conference on Antennas and Propagation. (EUCAP), Prague, Czech*, 2815–18.

Lee R. Q., Lee K. F., and Bobinchak J. (1987). Characteristics of a two-layer electromagnetically coupled rectangular patch antenna. *Electronics Letters*, 23, 1070–72.

Lee K. F., Luk K. M., Tong K. F., Shum S. M., Huynh T., and Lee R. Q. (1997). Experimental and simulation studies of the coaxially-fed U-slot rectangular patch antenna. *IEE Proceedings of the Microwaves Antennas and Propagations*, 144, 354–58.

Lee K. F., and Tong K. F. (2012). Microstrip patch antennas-basic characteristics and some recent advances. *Proceedings of the IEEE*, 100, 2169–80.

Lee M. W. K., Leung K. W., and Chow Y. L. (2015). Dual polarization slotted miniature wideband patch antenna. *IEEE Transactions on Antennas and Propagation*, 63, 353–57.

Lee K. F., Luk K. M., and Lai H. W. (2017). Microstrip patch antennas. *World Scientific*.

Leontovich, M. and Levin M. (1944). Towards a theory on the simulation of the oscillations in dipole antennas (in Russian). *Zhunal Teknicheskoi Fiziki*, 14, 481–506.

Lewin L. (1960). Radiation from discontinuities in strip-line. *Proceedings of the IRE Part C, Monographs*, 107, 163–70.

Li H., and Li Y. (2020). Mode compression method for wideband dipole antenna by dual-point capacitive loadings. *IEEE Transactions on Antennas and Propagation*, 68, 6424–28.

Li L., Zhang J., Lu W. J., Zhang W. H., and Zhu L. (2016). Dual-mode planar end-fire circularly polarized antenna. *IEEE International Workshop on Electromagnatics: Applications and Student Innovation Competition (iWEM), Nanjing, China*. 1–3.

Li Q., Lu W. J., Wang S. G., and Zhu L. (2017). Planar quasi-isotropic magnetic dipole antenna using fractional-order circular sector cavity resonant mode. *IEEE Access*, 5, 8515–25.

Li R., Guo Y.-X., Zhang B., and Du G. (2017). A miniaturized circularly polarized implantable annular-ring antenna. *IEEE Antennas and Wireless Propagation Letters*, 16, 2566–69.

Li R. L., Bushyager N. A., Laskar J., and Tentzeris M. M. (2005). Determination of reactance loading for circularly polarized circular loop antennas with a uniform traveling-wave current distribution. *IEEE Transactions on Antennas and Propagation*, 53, 3920–29.

Li S. J., Lu W. J., and Zhu L. (2021). Dual-band stacked patch antenna with wide e-plane beam width and stable gain at both bands. *Microwave and Optical Technology Letters*, 63, 1264–70.

Li W. A., Tu Z. H., Chu Q. X., and Wu X. H. (2015). Differential stepped-slot UWB antenna with common-mode suppression and dual sharp-selectivity notched bands. *IEEE Antennas and Wireless Propagation Letters*, 14, 1120–23.

Li X.-Q, Gao C., Lu W.-J., and Zhu H.-B. (2018). Preliminary studies of an offset-fed loop-dipole antenna for all-metal handsets. *International Workshop on Antenna Technology (iWAT'2018), Nanjing, China*, 1–3.

Li Y., Zhang Z., Feng Z., and Iskander M. F. (2011). Dual-mode loop antenna with compact feed for polarization diversity. *IEEE Antennas and Wireless Propagation Letters*, 10, 95–98.

Li Z., Shao Y., Li X. Q., and Lu W. J. (2019). A wideband triple-resonant stacked patch antenna. *IEEE International Wireless Symposium (IWS2019), Guangzhou, China*, 1–3.

Li Z. (2020). *Study on Multi-Resonant Stacked Circular Sector Patch Antennas.* Thesis of Master's Degree of Engineering, Nanjing University of Posts and Telecommunications.

Lian R., Wang Z., Yin Y., Wu J., and Song X. (2016). Design of a low-profile dual-polarized stepped slot antenna array for base station. *IEEE Antennas and Wireless Propagation Letters*, 15, 362–65.

Liang J., Chiau C. C., Chen X. et al. (2004). Printed circular disk monopole antenna for ultra-wideband applications. *Electronics Letters*, 40, 246–47.

Lindsay J. E. (1960). A circular loop antenna with nonuniform current distribution. *IRE Transactions on Antennas and Propagation*, AP-8, 439–41.

Liu C., Zhang Y., and Liu X. (2018). Circularly polarized implantable antenna for 915 MHz ISM-band far-field wireless power transmission. *IEEE Antennas and Wireless Propagation Letters*, 17, 373–76.

Liu C. N., Yuan C. Y., Lu W. J., and Zhu H. B. (2018). Conceptual design of wideband balanced circularly polarized dual-loop antenna. *International Workshop on Antenna Technology (iWAT'2018), Nanjing, China*, 1–3.

Liu C. N. (2019). *Study on Broadband Circularly Polarized Non-Uniform Square Loop Antenna (In Chinese).* Thesis of Master of Engineering, Nanjing University of Posts and Telecommunications.

Liu J. H., Xue Q., Wong H., Lai H. W., and Long Y. L. (2013). Design and analysis of a low-profile and broadband microstrip monopolar patch antenna. *IEEE Transactions on Antennas and Propagation*, 61, 11–18.

Liu N.-W., Zhu L., Choi W.-W., and Zhang J.-D. (2016). A novel differential-fed patch antenna on stepped-impedance resonator with enhanced bandwidth under dual-resonance. *IEEE Transactions on Antennas and Propagation*, 64, 4618–25.

Liu N. W., Zhu L., Choi W. W., and Zhang J. D. (2017). A low-profile aperture-coupled microstrip antenna with enhanced bandwidth under dual-resonance. *IEEE Transactions on Antennas and Propagation*, 65, 1055–62.

Liu N.-W., Zhu L., Choi W., and Zhang X. (2017). Wideband shorted patch antenna under radiation of dual-resonant modes. *IEEE Transactions on Antennas and Propagation*, 65, 2789–96.

Liu N. W., Zhu L., Fu G., and Liu Y. (2018). A low profile shorted-patch antenna with enhanced bandwidth and reduced H-plane cross-polarization. *IEEE Transactions on Antennas and Propagation*, 66, 5602–07.

Liu N. W., Zhu L., Choi W. W., and Fu G. (2019). A low-profile wideband aperture-fed microstrip antenna with improved radiation patterns. *IEEE Transactions on Antennas and Propagation*, 67, 562–67.

Liu S. B., Zhang F. S., and Zhang Y. X. (2019). Dual-band circular-polarization reconfigurable liquid dielectric resonator antenna. *International Journal of RF Microwave Computer-Aided Engineering*, 29, e21613.

Liu W., Chen Z. N., and Qing X. M. (2014). Metamaterial-based low-profile broadband mushroom antenna. *IEEE Transactions on Antennas and Propagation*, 62, 1165–72.

Liu X. Y., Wu Z. T., Fan Y., and Tentzeris E. M. (2016). A miniaturized CSRR loaded wide-beamwidth circularly polarized implantable antenna for subcutaneous real-time glucose monitoring. *IEEE Antennas and Wireless Propagation Letters*, 16, 577–80.

Lo T. K., Ho C. O., Hwang Y., Lam E. K.W. and Lee B. (1997). Miniature aperture-coupled microstrip antenna of very high directivity. *Electronics Letters*, 33, 9–10.

Lo Y. T., Solomon D., and Richards W. (1979). Theory and experiment on microstrip antennas. *IEEE Transactions on Antennas and Propagation*, AP-27, 137–45.

Long S. A. and Walton M. D. (1979). A dual-frequency stacked circular-disc antenna. *IEEE Transactions on Antennas and Propagation*, AP-27, 270–73.

Long S., McAllister M., and Shen L. (1983). The resonant cylindrical dielectric cavity antenna. *IEEE Transactions on Antennas and Propagation*, AP-31, 406–12.

Lu J. H., Tang C. L., and Wong K. L. (1998). Slot-coupled compact broadband circular microstrip antenna with chip-resistor and chip-capacitor loadings. *Microwave and Optical Technology Letters*, 18, 345–49.

Lu J. H., and Wang S. F. (2013). Planar broadband circularly polarized antenna with square slot for UHF RFID reader. *IEEE Transactions on Antennas and Propagation*, 61, 45–53.

Lu W. J., Cheng Y., and Zhu H. B. (2010). Design concept of a novel balanced ultra-wideband (UWB) antenna. *International Conference on Ubiquitous Wireless Broadband (ICUWB), Nanjing, China*. 1–4.

Lu W. J., Cheng C. H., and Zhu H. B. (2011). Wideband coplanar waveguide to edges-even broadside-coupled stripline transition. *Electronics Letters*, 47, 1286–87.

Lu W. J., Chen Y., Xu D. W. and Zhu H. B. (2011). Experimental investigation and comparison of super-wideband antipodal slot antennas. *International Journal of RF and Microwave Computer-Aided Engineering*, 21, 629–35.

Lu W. J., Bo Y. M. and Zhu H. B. (2012). A broadband transition design for a conductor-backed coplanar waveguide and a broadside coupled stripline. *IEEE Microwave and Wireless Components Letters*, 22, 10–12.

Lu W.-J., Bo Y.-M. and Zhu H.-B. (2012). Novel planar dual-band balanced antipodal slot-dipole composite antenna with reduced ground plane effect. *International Journal of RF and Microwave Computer-Aided Engineering*, 22, 319–28.

Lu W. J., Tong H., Bo Y. M. and Zhu H. B. (2013). Design and study of enhanced wideband transition between coplanar waveguide and broadside coupled stripline. *IET Microwaves Antennas and Propagation*, 7, 715–21.

Lu W. J., Zhang W. H., Tong K. F., and Zhu H. B. (2014). Planar wideband loop-dipole composite antenna. *IEEE Transactions on Antennas and Propagation*, 62, 2275–79.

Lu W. J., Liu G. M., Tong K. F., and Zhu H. B. (2014). Dual-band loop-dipole composite unidirectional antenna for broadband wireless communications. *IEEE Transactions on Antennas and Propagation*, 62, 2860–66.

Lu W. J., and Zhu L. (2015a). Wideband stub-loaded slotline antennas under multi-mode resonance operation. *IEEE Transactions on Antennas and Propagation*, 63, 818–23.

Lu W. J. and Zhu L. (2015b). Planar dual-mode wideband antenna using short-circuited-strips loaded slotline radiator: operation principle, design, and validation. *International Journal of RF and Microwave Computer-Aided Engineering*, 25, 573–81.

Lu W. J. and Zhu L. (2015c). A novel wideband slotline antenna with dual resonances: principle and design approach. *IEEE Antennas and Wireless Propagation Letters*, 14, 795–98.

Lu W. J., Shi J. W., Tong K. F., and Zhu H. B. (2015). Planar endfire circularly polarized antenna using combined magnetic dipoles. *IEEE Antennas and Wireless Propagation Letters*, 14, 1263–66.

Lu W. J., Li Q., Wang S. G., and Zhu L. (2017). Design approach to a novel dual-mode wideband circular sector patch antenna. *IEEE Transactions on Antennas and Propagation*, 65, 4980–90.

Lu W. J., Zhu L., Tam K. W., and Zhu H. B. (2017). Wideband dipole antenna using multi-mode resonance concept. *International Journal of Microwave and Wireless Technologies*, 9, 365–71.

Lu W. J., Li X. Q., Li Q., and Zhu L. (2018). Generalized design approach to compact wideband multi-resonant patch antennas. *International Journal of RF and Microwave Computer-Aided Engineering*, 28, e21481.

Lu W. J., Wang K., Gu S.-S., Zhu L., and Zhu H. B. (2019). Directivity enhancement of planar endfire circularly polarized antenna using V-shaped 1.5-wavelength dipoles. *IEEE Antennas and Wireless Propagation Letters*, 18, 1420–23.

Lu W. J., Yu J., and Zhu L. (2019). On the multi-resonant antennas: Theory, history and new development. *International Journal of RF and Microwave Computer-Aided Engineering*, 29, e21808.

Lu W. J., Yu J., and Zhu H. B. (2019). Research progress of the antenna technology for internet of things (in Chinese). *Telecommunications Science*, 2019, 124–35.

Lu X., Ni L., Jin S., Wen C.-K, and Lu W. -J. (2017). SDR implementation of a real-time testbed for future multi-antenna smartphone applications. *IEEE Access*, 5, 19761–72.

Lu Y. C., Yu M. J., and Lin Y. C. (2012). A single-fed slot-aperture hybrid antenna for broadband circular polarization operations. *Microwave and Optical Technology Letters*, 54, 412–15.

Ludwig A. (1973). The definition of cross polarization. *IEEE Transactions on Antennas and Propagation*, 21, 116–19.

Lui W. J., Cheng C. H., Cheng Y., and Zhu H. (2005a). Frequency notched ultra-wideband microstrip slot antenna with fractal tuning stub. *Electronics Letters*, 41, 294–96.

Lui W. J., Cheng C. H., Cheng Y., and Zhu H. (2005b). Frequency notched printed slot antenna with parasitic open-circuit stub. *Electronics Letters*, 41, 1094–95.

Lui W. J., Cheng C. H., Cheng Y. and Zhu H. B. (2005c). A novel broadband multislot antenna fed by microstrip line. *Microwave and Optical Technology Letters*, 45, 55–57.

Lui W. J., Cheng C. H. and Zhu H. B. (2006). Compact frequency notched ultra-wideband fractal printed slot antenna. *IEEE Microwave and Wireless Components Letters*, 16, 224–26.

Lui W.-J., Cheng C. H., and Zhu H.-B. (2007a). Experimental investigation on novel tapered microstrip slot antenna for ultra-wideband applications. *IET Microwaves Antennas and Propagation*, 1480–87.

Lui W. J., Cheng C. H. and Zhu H. B. (2007b). Improved frequency notched ultra-wideband slot antenna using square ring resonator. *IEEE Transactions on Antennas and Propagation*, 55, 2445–50.

Lui W. J. (2008). Ultra-wideband folded loop antenna fed by coplanar waveguide. *Microwave and Optical Technology Letters*, 50, 3075–77.

Luk K. M., Mak C. L., Chow Y. L., and Lee K. F. (1998). Broadband microstrip patch antenna. *Electronics Letters*, 34, 1442–43.

Luk K. M., and Leung K. W. (2003). *Dielectric Resonator Antennas*. UK: Research Studies Press.

Luk K. M. and Wong H. (2006). A new wideband unidirectional antenna element. *International Journal of Microwave and Optical Technology*, 1, 35–44.

Luk K. and Wu B. (2012). The magnetoelectric dipole—A wideband antenna for base stations in mobile communications. *Proceedings of the IEEE*, 100, 2297–2307.

Luo T., Feng B., and Q Zeng. (2017). A novel magneto-electric dipole antenna with wide H-plane triple-linear polarization and frequency reconfiguration for 5G application. *International Symposium on Electromagnetic Compatibility, (EMC-Beijing), Beijing, China*, 1–3.

Luo Y., Chen Z. N., and Ma K. (2019). Enhanced bandwidth and directivity of a dual-mode compressed high-order mode stub-loaded dipole using characteristic mode analysis. *IEEE Transactions on Antennas and Propagation*, 67, 1922–25.

Mao C. X., Gao S., Wang Y., Luo Q., and Chu Q. X. (2017). A shared-aperture dual-band dual-polarized filtering-antenna-array with improved frequency response. *IEEE Transactions on Antennas and Propagation*, 65, 1836–44.

Marcuvitz N. (1956). On field representations in terms of leaky modes or eigen-modes. *IRE Transactions on Antennas and Propagation*, AP-4, 192–94.

Mayes P. E., Warren W. T., and Wiesenmeyer F. M. (1971). The monopole-slot: a broadband, unidirectional antenna. *IEEE International Antennas and Propagation Symposium Digest, Los Angeles, CA, USA*, 109–12.

Mcilvena J., and Kernweis N. (1979). Modified circular microstrip antenna elements. *Electronics Letters*, 15, 207–08.

Mehta A., and Mirshekar-Syahkal D. (2007). Pattern steerable square loop antenna. *Electronics Letters*, 43, 491–93.

Menzel W. (1978). A new travelling wave antenna in microstrip. *European Microwave Conference Paris, France*, 302–06.

Morishita H., Hirasawa K., and Nagao T. (1998). Circularly polarised wire antenna with a dual rhombic loop. *IEE Proceedings of Microwaves Antennas and Propagations*, 145, 219–24.

Motevasselian A. and Whittow W. G. (2017). Miniaturization of a circular patch microstrip antenna using an arc projection. *IEEE Antennas and Wireless Propagation Letters*, 16, 517–20.

Munson R. E. (1974). Conformal microstrip antennas and microstrip phased arrays. *IEEE Transactions on Antennas and Propagation*, AP-22, 74–78.

Murakarni Y., Yoshida A., Ida F., and Nakamura T. (1996). Rectangular loop antenna for circular polarization. *Electronics and Communications in Japan, Part 1*, 79, 42–51.

Mushiake Y. (1996). *Self-Complementary Antennas: Principle of Self-Complementarity for Constant Impedance*. Berlin: Springer.

Nakano H. (1998). Loop antenna for radiating circularly polarized waves. US Patent 5, 838, 283.

Nakano H., Fujimori K., and Yamaguchi J. (2000). A low-profile conical beam loop antenna with an electromagnetically coupled feed system. *IEEE Transactions on Antennas and Propagation*, 48, 1864–66.

Nakano H., Yamamoto Y., Seto M., Hitosugi K., and Yamauchi J. (2004). A half-moon antenna. *IEEE Transactions on Antennas and Propagation*, 52, 3237–44.

Namin F., Spence T. G., Werner D. H., and Semouchkina E. (2010). Broadband, miniaturized stacked-patch antennas for L-band operation based on magneto-dielectric substrates. *IEEE Transactions on Antennas and Propagation*, 58, 2817–22.

Navarro-Mendez D. V., Carrera-Suarez L. F., Sanchez-Escuderos D., Cabedo-Fabres M., Baquero-Escudero M., Gallo M., and Zamberlan D. (2017). Wideband double monopole for mobile, WLAN, and C2C services in vehicular applications. *IEEE Antennas and Wireless Propagation Letters*, 16, 16–19.

Niemi T., Alitalo P., Karilainen A. O., and Tretyakov S. A. (2012). Electrically small Huygens source antenna for linear polarization. *IET Microwaves Antennas and Propagation*, 6, 735–39.

Noghanian S. and Shafai L. (1998). Control of microstrip antenna radiation characteristics by ground plane size and shape. *IEE Proceedings of Microwaves Antennas and Propagations*, 145, 207–12.

Oliner A. A. (1957a). The impedance properties of narrow radiating slots in the broad face of rectangular waveguide, Part I: Theory. *IRE Transactions on Antennas and Propagation*, AP-5, 4–11.

Oliner A. A. (1957b). The impedance properties of narrow radiating slots in the broad face of rectangular waveguide, Part II: Comparison with measurement. *IRE Transactions on Antennas and Propagation*, AP-5, 12–20.

Ooi B.L., Qin S., and Leong M. S. (2002). Novel design of broad-band stacked patch antenna. *IEEE Transactions on Antennas and Propagation*, 50, 1391–95.

Osklang P., and Phongcharoenpanich C. (2016). Broadband planar dipole array antenna with double C-shaped slit elements for digital TV broadcasting transmission. *International Journal of RF and Microwave Computer-aided Engineering*, 26, 466–78.

Pan C.-X., Lu W.-J., Jia W.-Q., and Zhu L. (2021). Triple-resonant wideband 1.5-wavelength sectorial dipole antenna. *International Journal of RF and Microwave Computer-Aided Engineering*, 31, e22728.

Pan G., Li Y., Zhang Z., and Feng Z. (2014). A compact wideband slot-loop hybrid antenna with a monopole feed. *IEEE Transactions on Antennas and Propagation*, 62, 3864–68.

Pan Z., Lin W. and Chu Q. (2014). Compact wide-beam circularly-polarized microstrip antenna with a parasitic ring for CNSS application. *IEEE Transactions on Antennas and Propagation*, 62, 2847–50.

Parihar M., Basu A., and Koul S. (2011). Efficient spurious rejection and null steering using slot antennas. *IEEE Antennas and Wireless Propagation Letters*, 10, 207–10.

Pazin L., and Leviatan Y. (2016). On the matching characteristics of a rectangular slot located near the edge of a finite-size ground plane. *Journal of Electromagnetic Waves and Applications*, 30, 579–88.

Podilchak S. K., Murdock A. P., and Antar Y. M. M. (2017). Compact, microstrip-based folded-shorted patches PCB antennas for use on microsatellites. *IEEE Antennas and Propagation Magazine*, 59, 88–95.

Ponchak G. E. (2016). Slotline switch based on a lattice circuit. *IEEE Microwave and Wireless Components Letters*, 26, 43–45.

Pozar D. M. (1985). A microstrip antenna aperture coupled to a microstrip line. *Electronics Letters*, 21, 49–50.

Pozar D. M., and Croq F. (1991). Millimeter-wave design of wide-band aperture-coupled stacked microstrip antennas. *IEEE Transactions on Antennas and Propagation*, 39, 1770–76.

Pozar D. M. (1992). Microstrip antennas. *Proceedings of the IEEE*, 80, 79–91.

Prasad S. M., and Das B. N. (1970). A circular loop antenna with traveling-wave current distribution. *IEEE Transactions on Antennas and Propagation*, AP-18, 278–80.

Qian J. F., Chen F. C., and Chu Q. X. (2018). A novel tri-band patch antenna with broadside radiation and its application to filtering antenna. *IEEE Transactions on Antennas and Propagation*, 66, 5580–85.

Queck T., Schünemann B., Radusch I., and Meinel C. (2008). Realistic simulation of V2X communication scenarios. *IEEE Asia-Pacific Services Computing Conference, Yilan, Taiwan*, 1623–27.

Quintero G., Zürcher J. F., and Skrivervik A. K. (2011). System fidelity factor: A new method for comparing UWB antennas. *IEEE Transactions on Antennas and Propagation*, 59, 2502–12.

Rayleigh L. (1897). On the passage of electric waves through tubes, or the vibration of dielectric cylinders. *Philosophical Magazine*, 43, 125–32.

Richards W., Lo Y. T., and Harrison D. D. (1981). An improved theory for microstrip antennas and applications. *IEEE Transactions on Antennas and Propagation*, AP-29, 38–46.

Richards W., Ou J. D., and Long S. A. (1984). A theoretical and experimental investigation of annular, annular sector, and circular sector microstrip antennas. *IEEE Transactions on Antennas and Propagation*, 32, 991–94.

Ruan X., and Chan C. H. (2020). An endfire circularly polarized complementary antenna array for 5G applications. *IEEE Transactions on Antennas and Propagation*, 68, 266–74.

Rumsey V. H. (1957). Frequency independent antennas. *IRE International Convention Record*, 5:114–18.

Schantz H. G. (2003). Bottom fed planar elliptical UWB antennas. *IEEE Conference on Ultra Wideband Systems and Technologies, Reston, VA, USA*, 219–23.

Schantz H. G., Wolenec G., and Myszka E. M. (2003). Frequency notched UWB antennas. *IEEE Conference on Ultra Wideband Systems and Technologies*, 214–218.

Schantz H. G. (2005).*The Art and Science of Ultrawideband Antennas* (First Edition). Boston, MA: Artech House.

Schantz H. (2015). *The Art and Science of Ultrawideband Antennas* (Second Edition), Chapter 4. Boston/London: Artech House.

Schelkunoff S. A. (1941). Theory of antennas of arbitrary size and shape. *Proceedings of the IRE*, 29, 493–521.

Schelkunoff S. A. (1945). Concerning Hallén's integral equation for cylindrical antennas. *Proceedings of the IRE*, 33, 872–78.

Schelkunoff S. A., and Friis H. T. (1952). *Antennas: Theory and Practice*. New York: John Wiley and Son's.

Samadi T. M. M., Hassani H. R. and Nezhad S. M. A. 2011. UWB printed slot antenna with Bluetooth and dual notch bands. *IEEE Antennas and Wireless Propagation Letters*, 10, 255–58.

Seo D. G., Kim J. H., Ahn S. H., and Lee W. S. (2018). A 915 MHz dual polarized meandered dipole antenna with dual resonance. *International Symposium on Antennas and Propagation (ISAP), Busan, Korea*, 1–2.

Sharma S. K., Shafai L., and Jacob N. (2004). Investigations of wide band microstrip slot antenna. *IEEE Transactions on Antennas and Propagation*, 52, 865–72.

Shao Y., Li X. Q., Lu W.-J., and Zhu H.-B. (2019a). Wideband dual-resonant fixed-beam high gain patch antenna array. *International Conference on Microwave and Millimeter Wave Technology (ICMMT2019), Guangzhou, China*, 1–3.

Shao Y., Li Z., Yu J., and Lu W. J. (2019b). Pin-loaded dual-resonant high gain patch antenna array with extremely thin profile. *The Asia-Pacific Conference on Antennas and Propagation (APCAP 2019), Incheon, South Korea*, 1–3.

Shao Y., (2020), Study on fixed beam high gain multiple resonant sector patch antenna array (in Chinese). Master of Engineering Degree's Thesis, Nanjing University of Posts and Telecommunications, China.

She J., Gao C., Yu Y., Cui P. F., Lu W.-J., Jin S., and Zhu H. B. (2017). Measurements of massive MIMO channel in real environment with 8-antenna handset. *International Conference on Wireless Communications and Signal Processing (WCSP'2017), Nanjing, China*, 1–4.

She J., Lu W.-J., Liu Y., Cui P.-F., and Zhu H. B. (2019). An experimental massive MIMO channel matrix model for hand-held scenarios. *IEEE Access*, 7, 33881–87.

Shen C., Lu W. J., and Zhu L. (2019). Planar self-balanced magnetic dipole antenna with wide beamwidth characteristic. *IEEE Transactions on Antennas and Propagation*, 67, 4860–65.

Sherman J. B. (1944). Circular loop antennas at ultra-high frequencies. *Proceedings of the IRE*, 32, 534–37.

Shin H., Jiang S., Yang J., Kim H., and Kim H. (2020). A new ultra-wideband miniaturized antenna with a double-branch radiator. *Microwave and Optical Technology Letters*, 62, 2085–89.

Shin J. and Schaubert D. H. (1999). A parameter study of stripline-fed Vivaldi notch-antenna arrays. *IEEE Transactions on Antennas and Propagation*, 47, 879–86.

Siddiqui J. Y., Saha C., and Antar Y. M. M. (2015). Compact dual-SRR-loaded UWB monopole antenna with dual frequency and wideband notch characteristics. *IEEE Antennas and Wireless Propagation Letters*, 14, 100–03.

Singh D., Gardner P. and Hall P.S. (1997). Miniaturised microstrip antenna for MMIC applications. *Electronics Letters*, 33, 1830–31.

Simpson T. (2006). The schelkunoff-friis dipole: the simplest antenna of all. *IEEE Antennas and Propagation Magazine*, 48, 48–53.

Sommers D. J. (1955). Slot array employing photoetched tri-plate transmission lines. *IRE Transactions on Microwave Theory and Techniques*, MTT-3, 157–62.

Stephenson B. T., and Walter C. H. (1955). Endfire slot antennas. *IRE Transactions on Antennas and Propagation*, AP-3, 81–86.

Sun D., and You L. (2010). A broadband impedance matching method for proximity-coupled microstrip antenna. *IEEE Transactions on Antennas and Propagation*, 58, 1392–97.

Sun G. H., Wong S. W., and Wong H. (2017). A broadband antenna array using full-wave dipole. *IEEE Access*, 5, 13054–61.

Sun Y. X., Leung K. W., and Ren J. (2018). Dual-band circularly polarized antenna with wide axial ratio beamwidths for upper hemispherical coverage. *IEEE Access*, 6, 58132–138.

Syrytsin I., Zhang S., Pedersen G. F., and Ying Z. (2018). User effects on the circular polarization of 5G mobile terminal antennas. *IEEE Transactions on Antennas and Propagation*, 66, 4906–11.

Sze J. and Wong K. (2001). Bandwidth enhancement of a microstrip-line-fed printed wide slot antenna. *IEEE Transactions on Antennas and Propagation*, 49, 1020–24.

Sze J. Y., Wong K. L. and Huang C. C. (2003). Coplanar waveguide-fed square slot antenna for broadband circularly polarized radiation. *IEEE Transactions on Antennas and Propagation*, 51, 2141–44.

Sze J. Y., Hsu C. I. G., Chen Z. W. and Chang C. C. (2010). Broadband CPW-fed circularly polarized square slot antenna with lightening-shaped feed-line and inverted-L grounded strips. *IEEE Transactions on Antennas and Propagation*, 58, 973–77.

Ta S. X., Choo H., Park I., and Ziolkowski R. W. (2013). Multi-band, wide-beam, circularly polarized, crossed, asymmetrically barbed dipole antennas for GPS applications. *IEEE Transactions on Antennas and Propagation*, 61, 5771–75.

Tai C. T. (1948). Coupled antennas. *Proceedings of the IRE*, 36, 487–500.

Tang M. C., Wu Z., Shi T., and Ziolkowski R. W. (2019). Dual-band, linearly polarized, electrically small Huygens dipole antennas. *IEEE Transactions on Antennas and Propagation*, 67, 37–47.

Targonski S. D., Waterhouse R. B., and Pozar D. M. (1998). Design of wideband aperture-stacked patch microstrip antennas. *IEEE Transactions on Antennas and Propagation*, 46, 1245–51.

Tefiku F. and Grimes C. A. (2000). Design of broad-band and dual-band antennas comprised of series-fed printed-strip dipole pairs. *IEEE Transactions on Antennas and Propagation*, 48, 895–900.

Thumvichit A., Takano T., and Kamata Y. (2007). Characteristics verification of a half-wave dipole very close to a conducting plane with excellent impedance matching. *IEEE Transactions on Antennas and Propagation*, 55, 53–58.

Tsai L. C. (2014). Bandwidth enhancement in coplanar waveguide-fed slot antennas designed for wideband code division multiple access/wireless local area network/worldwide interoperability for microwave access applications. *IET Microwaves Antennas and Propagation*, 8, 724–29.

Uthansakul M. and Bialkowski M. E. (2006). Wideband beam and null steering using a rectangular array of planar monopoles. *IEEE Microwave and Wireless Components Letters*, 16, 116–18.

Vandenbosch G A. E., and Van de Capelle A. R. (1994). Study of the capacitively fed microstrip antenna element. *IEEE Transactions on Antennas and Propagation*, 42, 1648–52.

Vaughan R. G. (1988). Two-port higher mode circular microstrip antennas. *IEEE Transactions on Antennas and Propagation*, 36, 309–21.

Vorobyov A. V., Zijderfeld J. H., Yarovoy A. G., and Ligthart L. P. (2005). Impact common mode currents on miniaturized UWB antenna performance. *Proceedings of the IEEE European Conference on Wireless Technology, Paris, France*, 285–88.

Wang D., Wong H., and Chan C. H. (2011). Miniaturized circularly polarized patch antenna by substrate integrated irregular ground. *IEEE International Symposium on Antennas and Propagation, Spokane, WA, USA*, 1875–77.

Wang D. -D., Yuan C. Y., Lu W. J., and Zhu H. B. (2017a). Conceptual design of a dual-band circularly polarized square loop antenna under even-mode resonance. *6th Asia-Pacific Conference on Antennas and Propagation (APCAP), Xi'an, China*, 1–3.

Wang H., Zhou D., Xue L., Gao S., and Xu H. (2017b). Modal analysis and excitation of wideband slot antennas. *IET Microwaves Antennas and Propagation*, 11, 1889–93.

Wang J.-Y, Lu W.-J., Zhang W.-L., and Zhu L. (2021). Balanced low-profile wide beamwidth circularly polarized stacked loop antenna. *International Journal of RF and Microwave Computer-Aided Engineering*, 32, e22848.

Wang S., Park J., Shim H., and Choo H. (2020). Design of a wideband coupled feed dipole antenna for PCL array systems. *Journal of Electrical Engineering and Technology*, 15, 2251–58.

Wang S. G., Lu W. J., Guo C. R., and Zhu L. (2017c). Wideband slotline antenna with a frequency-spatial steerable notch-band in radiation gain. *Electronics Letters*, 53, 650–52.

Wang Z., Wu J., Yin Y. and Liu X. (2014). A broadband dual-element folded dipole antenna with a reflector. *IEEE Antennas and Wireless Propagation Letters*, 13, 750–53.

Waterhous R. B. (1995). Small microstrip patch antenna. *Electronics Letters*, 31, 604–05.

Waterhouse R. B. (1999). Design of probe-fed stacked patches. *IEEE Transactions on Antennas and Propagation*, 47, 1780–84.

Waterhouse R. B. (2013). *Microstrip Patch Antennas: A Designer's Guide*. Springer Science & Business Media.

Watkins J. (1969). Circular resonant structures in microstrip. *Electronics Letters*, 5, 524–25.

Weedon W. H., Payne W. J., and Rebeiz G.M. (2001). MEMS-switched reconfigurable antennas. *IEEE Antennas and Propagation Society International Symposium Digest, Boston, MA, USA*, 654–57.

Wen D. L., and Hao Y. (2016). A wideband T-shaped slot antenna and its MIMO application. *IEEE International Symposium on Antennas and Propagation (APSURSI), Puerto Rico, USA*, 1–2.

Wen D. -L., Zheng D.-Z., and Chu Q. X. (2017). Wideband differentially fed dual-polarized antenna with stable radiation pattern for base stations. *IEEE Transactions on Antennas and Propagation*, 65, 2248–55.

Wen D. L., Hao Y., Wang H.Y. and Zhou H. (2017). Design of a wideband antenna with stable omnidirectional radiation pattern using the theory of characteristic modes. *IEEE Transactions on Antennas and Propagation*, 65, 2671–76.

Werner D. H. (2000). Near-field and far-field expansions for traveling-wave circular loop antennas. *Progress in Electromagnetics Research-Pier*, 28, 29–42.

Wheeler H. A. (1947). A helical antenna for circular polarization. *Proceedings of the IRE*, 35, 1484–88.

Wheeler H. A. (1950). Directive loop antenna. US Patent 2, 518, 736.

Wolff I. (1972). Microstrip bandpass filter using degenerate modes of a microstrip ring resonator. *Electronics Letters*, 8, 302–03.

Wolosinski G., Fusco V., and Rulikowski P. (2019). High-performance balun for a dual-polarised dipole antenna. *IET Microwaves Antennas and Propagation*, 13, 346–51.

Wong H., Mak K. M., and Luk K. M. (2008). Wideband shorted bowtie patch antenna with electric dipole. *IEEE Transactions on Antennas and Propagation*, 56, 2098–2101.

Wong H., So K. K., and Gao X. (2016). Bandwidth enhancement of a monopolar patch antenna with V-shaped slot for car-to-car and WLAN communications. *IEEE Transactions on Vehicular Technology*, 65, 1130–36.

Wong K. K., Tong K. F., Zhang Y., and Zheng Z. (2020). Fluid antenna system for 6G: When Bruce Lee inspires wireless communications. *Electronics Letters*, 56, 1288–90.

Wong K. L. (2002). *Compact and Broadband Microstrip Antenna*. Chapter 3. John Wiley and Son's, New York.

Wong T. P. and Luk K. M. (2005). A wide bandwidth and wide beamwidth CDMA/GSM base station antenna array with low backlobe radiation. *IEEE Transactions on Vehicular Technology*, 54, 903–09.

Wu H., Lu W. J., Shen C., and Zhu L. (2020). Wide beamwidth planar self-balanced magnetic dipole antenna with enhanced front-to-back ratio. *International Journal of RF and Microwave Computer-aided Engineering*, 30, e22171.

Wu Q. S., and Zhu L. (2018). Co-design of a wideband circularly polarized filtering patch antenna with three minima in axial ratio response. *IEEE Transactions on Antennas and Propagation*, 66, 5022–30.

Wu Q. S., Zhang X., and Zhu L. (2018). A wideband circularly polarized patch antenna with enhanced axial ratio bandwidth via co-design of feeding network. *IEEE Transactions on Antennas and Propagation*, 66, 4996–5003.

Wu T. T. (1962). Theory of the thin circular loop antenna. *Journal of Mathematical Physics*, 3, 1301–04.

Wu X. T., Lu W. J., Xu J., Tong K. F. and Zhu H. B. (2015). Loop-monopole composite antenna for dual-band wireless communications. *IEEE Antennas and Wireless Propagation Letters*, 14, 293–96.

Wu Z. F., Lu W. J., Yu J. and Zhu L. (2020). Wideband null frequency scanning circular sector patch antenna under triple-resonance. *IEEE Transactions on Antennas and Propagation*, 68, 7266–74.

Wu Z. F., Yu J., and Lu W. J. (2020). Investigations of multi-resonant wideband null frequency scanning microstrip patch antennas. *Asia-Pacific Microwave Conference*, 1–3.

Xiao S. Q., Wang B. Z., Shao W., and Zhang Y. (2005). Bandwidth-enhancing ultralow-profile compact patch antenna. *IEEE Transactions on Antennas and Propagation*, 53, 3443–47.

Xing L., Xu Q., Zhu J., Zhao Y., Alja'afreh S., Song C., and Huang Y. (2021). A high-efficiency wideband frequency-reconfigurable water antenna with a liquid control system: Usage for VHF and UHF applications. *IEEE Antennas and Propagation Magazine*, 63, 61–70.

Xu H., Wang H.-Y., and Gao S. (2016). A compact and low profile loop antenna with six resonant modes for LTE smartphone. *IEEE Transactions on Antennas and Propagation*, 64, 3743–51.

Xu H., Gao S., Zhou H., Wang H., and Cheng Y. (2019). A highly integrated MIMO antenna unit: Differential/common mode design. *IEEE Transactions on Antennas and Propagation*, 67, 6724–34.

Xu J., Lu W.-J., Wu X.-T., Bo Y.-M., Zhu L., and Zhu H.-B. (2015). Novel offset-fed dual-band aperture-dipole composite antenna: operating principle and design approach. *International Journal of RF and Microwave Computer-aided Engineering*, 25, 382–93.

Xu L. J., Bo Y.-M., Lu W.-J., Zhu L. et al. (2019). Circularly polarized annular ring antenna with wide axial-ratio bandwidth for biomedical applications. *IEEE Access*, 7, 59999–60009.

Xu L. J., Jin X.-Y., Hua D., Lu W. -J., and Duan Z. (2020). Realization of circular polarization and gain enhancement for implantable antenna. *IEEE Access*, 8, 16857–64.

Xu K. D., Li D., Liu Y., and Liu Q.H. (2018). Printed quasi-Yagi antennas using double dipoles and stub-loaded technique for multi-band and broadband applications. *IEEE Access*, 6, 31695–702.

Xu L., Lu W. J., Yuan C. Y., and Zhu L. (2019). Dual circularly polarized loop antenna using a pair of resonant even modes. *International Journal of RF and Microwave Computer-Aided Engineering*, 29, e21703.

Xue B., You M., Lu W. J., and Zhu L. (2016). Planar endfire circularly polarized antenna using concentric annular sector complementary dipoles. *International Journal of RF and Microwave Computer-aided Engineering*, 26, 829–38.

Yagi H. (1928). Beam transmission of ultra short waves. *Proceedings of the IRE*, 16, 715–40.

Yan S. and Vandenbosch Guy A. E. (2016). Meta-loaded circular sector patch antenna. *Progress in Electromagnetics Research-Pier*, 156, 37–46.

Yang F., Zhang X. X., Ye X., and Rahmat-Samii Y. (2001). Wide-band E-shaped patch antennas for wireless communications. *IEEE Transactions on Antennas and Propagation*, 49, 1094–1100.

Yang G., Chu Q. X., Tu Z. and Wang Y. (2012). Compact printed dipole antenna with integrated wideband balun for UWB application. *International Conference on Microwave and Millimeter Wave Technology (ICMMT)*, Shenzhen, China, 1–4.

Yang G., Li J., Wei D., Zhou S., and Yang J. (2018). Broadening the beam-width of microstrip antenna by the induced vertical currents. *IET Microwaves Antennas and Propagation*, 12, 190–94.

Yang G., Li J., Yang J., and Zhou S. (2018). A wide beamwidth and wideband magnetoelectric dipole antenna. *IEEE Transactions on Antennas and Propagation*, 66, 6724–33.

Yang H. Q., You M., Lu W. J., Zhu L. and Zhu H. B. (2018). Envisioning an endfire circularly polarized antenna: presenting a planar antenna with a wide beamwidth and enhanced front-to-back ratio. *IEEE Antennas and Propagation Magazine*, 60, 70–79.

Yashchyshyn Y., and Starszuk G. (2005). Investigation of a simple four-element null steering antenna array. *IEE Proceedings of the Microwaves Antennas and Propagations*, 92–96.

Yin J., Wu Q., Yu C., Wang H., and Hong W. (2019). Broadband symmetrical E-shaped patch antenna with multimode resonance for 5G millimeter-wave applications. *IEEE Transactions on Antennas and Propagation*, 67, 4474–83.

Yoshimura Y. (1972). A microstripline slot antenna. *IEEE Transactions on Microwave Theory and Techniques*, 20, 760–62.

Yong S., and Bernhard J. T. (2013). Reconfigurable null scanning antenna with three dimensional null steer. *IEEE Transactions on Antennas and Propagation*, 61, 1063–70.

You M., Lu W. J., Xue B., Zhu L., and Zhu H. B. (2016). A novel planar endfire circularly polarized antenna with wide axial ratio beamwidth and wide impedance bandwidth. *IEEE Transactions on Antennas and Propagation*, 64, 4554–59.

You M., Lu W. J., Cheng Y., and Zhang W. H. (2016). Preliminary studies on the coupled antennas and feedlines using natural boundary conditions. *IEEE International Conference on Ubiquitous Wireless Broadband (ICUWB)*, Nanjing, China, 1–4.

Yousefi L. and Ramahi O.M. (2010). Miniaturised antennas using artificial magnetic materials with fractal Hilbert inclusions. *Electronics Letters*, 46, 816–17.

Yu J., and Lu W.-J. (2019). Design approach to dual-resonant, very low-profile circular sector patch antennas. *International Conference on Microwave and Millimeter Wave and Technology (ICMMT2019), Guangzhou, China.*

Yu J., Lu W. J., Cheng Y., and Zhu L. (2020). Tilted circularly polarized beam microstrip antenna with miniaturized circular sector patch under wideband dual-mode resonance. *IEEE Transactions on Antennas and Propagation*, 68, 6580–90.

Yu Y., Cui P. F., She J., Liu Y., Yang X., Lu W. J., Jin S., Zhu H. B. (2016). Measurement and empirical modeling of massive MIMO channel matrix in real indoor environment. *International Conference on Wireless Communications and Signal Processing (WCSP'2016), Yangzhou, China*, 1–4.

Yuan C. Y., Lu W. J., Gao C., and Zhu L. (2018). Balanced circularly polarized square loop antenna under even-mode resonance. *International Journal of RF and Microwave Computer-aided Engineering*, 28, e21280.

Zang Y., Zhai H., Xi L., and Li L. (2019). A compact microstrip antenna with enhanced bandwidth and ultra-wideband harmonic suppression. *IEEE Transactions on Antennas and Propagation*, 67, 1969–74.

Zhang J., Lu W. J., Li L., Zhu L., and Zhu H. B. (2016). Wideband dual-mode planar endfire antenna with circular polarization. *Electronics Letters*, 52, 1000–01.

Zhang J. (2017). Study on planar circularly polarized complementary dipole antenna (in Chinese). Master of Engineering. Degree's Thesis, Chapter 5, Nanjing University of Posts and Telecommunications, China.

Zhang W. H., Lu W. J., and Tam K. W. (2016). A planar end-fire circularly polarized complementary antenna with beam in parallel with its plane. *IEEE Transactions on Antennas and Propagation*, 64, 1146–52.

Zhang W. H., Cheong P., Lu W. J., and Tam K. W. (2018). Planar endfire circularly polarized antenna for low profile handheld RFID reader. *IEEE Journal of Radio Frequency Identification*, 2, 15–22.

Zhang W. L., Wang D. D., Xu L., Lu W. J., and Zhu H. B. (2019). Improved design of a dual-band circularly polarized square loop antenna under even-mode resonance. *International Conference on Microwave and Millimeter Wave and Technology (ICMMT2019), Guangzhou, China,* 1–3.

Zhang W. L. (2020). *Study on Circularly Polarized Non-Uniform Circular Loop Antenna under Even-Mode Resonance (in Chinese)*. Thesis of Master of Engineering. Nanjing University of Posts and Telecommunications.

Zhang W. Q., Li Y., Zhou Z. H., and Zhang Z. J. (2020). Dual-mode compression of dipole antenna by loading electrically small loop resonator. *IEEE Transactions on Antennas and Propagation*, 68, 3243–47.

Zhang W. X. (1982). *Partial Differential Equations in Radio Techniques (in Chinese)*, Chapter 10, Section 10.4. Beijing: Publishing House of National Defense Industry.

Zhang X. and Zhu L. (2016a). Patch antennas with loading of a pair of shorting pins toward flexible impedance matching and low cross-polarization. *IEEE Transactions on Antennas and Propagation*, 64, 1226–33.

Zhang X., and Zhu L. (2016b). High-gain circularly polarized microstrip patch antenna with loading of shorting pins. *IEEE Transactions on Antennas and Propagation*, 64, 2172–78.

Zhang X. and Zhu L. (2016c). Gain-enhanced patch antennas with loading of shorting pins. *IEEE Transactions on Antennas and Propagation*, 64, 3310–18.

Zhang X., Zhu L., and Liu N.-W. (2017). Pin-loaded circularly-polarized patch antennas with wide 3-dB axial ratio beamwidth. *IEEE Transactions on Antennas and Propagation*, 65, 521–28.

Zhang X. and Zhu L. (2017). Gain-enhanced patch antenna without enlarged size via loading of slot and shorting pins. *IEEE Transactions on Antennas and Propagation*, 65, 5702–09.

Zhang X. and Zhu L. (2018a). Side-lobe-reduced and gain-enhanced square patch antennas with adjustable beamwidth under TM03 mode operation. *IEEE Transactions on Antennas and Propagation*, 66, 1704–13.

Zhang X. and Zhu L. (2018b). Dual-band high-gain differentially fed circular patch antenna working in TM11 and TM12 modes. *IEEE Transactions on Antennas and Propagation*, 66, 3160–65.

Zhang X., Zhu L., Liu N. W., and Xie D. (2018). Pin-loaded circularly-polarised patch antenna with sharpened gain roll-off rate and widened 3-dB axial ratio beamwidth. *IET Microwaves Antennas and Propagation*, 12, 1247–54.

Zhang X., Wu Q. S., Zhu L., Huang G.-L., and Yuan T. (2019). Resonator-fed wideband and high-gain patch antenna with enhanced selectivity and reduced cross-polarization. *IEEE Access*, 7, 49918–27.

Zhang X. Y., Duan W., and Pan Y. M. (2015). High-gain filtering patch antenna without extra circuit. *IEEE Transactions on Antennas and Propagation*, 63, 5883–88.

Zhang Y., Wei K., Zhang Z., Li Y., and Feng Z. (2015). A compact dual-mode metamaterial-based loop antenna for pattern diversity. *IEEE Antennas and Wireless Propagation Letters*, 14, 394–97.

Zhang Y. P. (2007). Design and experiment on differentially-driven microstrip antennas. *IEEE Transactions on Antennas and Propagation*, 55, 2701–08.

Zhang Z. S., Chen Y., Guo C. R., Lu W. J., and Zhu L. (2016). Conceptual design of a circularly polarized non-uniform square loop antenna. *IEEE International Workshop on Electromagnetics: Applications and Student Innovation Competition (iWEM), Nanjing, China*, 1–3.

Zhao Z. B., Lu W.-J., Zhu L., and Yu J. (2021). Wideband wide beamwidth full-wavelength sectorial dipole antenna under dual-mode resonance. *IEEE Transactions on Antennas and Propagation*, 69, 14–24.

Zheng D.-Z., and Chu Q. X. (2017). A wideband dual-polarized antenna with two independently controllable resonant modes and its array for base-station applications. *IEEE Antennas and Wireless Propagation Letters*, 16, 2014–17.

Zhong S. S., and Lo Y. T. (1983). Single-element rectangular microstrip antenna for dual-frequency operation. *Electronics Letters*, 19, 298–300.

Zhu J., and Roy S. (2003). MAC for dedicated short range communications in intelligent transport system. *IEEE Communications Magazine*, 41, 60–67.

Zhu L., Fu R., and Wu K. L. (2003). A novel broadband microstrip-fed wide slot antenna with double rejection zeros. *IEEE Antennas and Wireless Propagation Letters*, 2, 194–96.

Zhu L., Sun S., and Menzel W., (2005). Ultra-wideband (UWB) bandpass filters using multiple-mode resonator. *IEEE Microwave and Wireless Components Letters*, 15, 796–98.

Index

Ampère's right-hand grip rule 6, 14
amplitude modulation (AM) 113
angular bisector 16, 17, 43, 212
aperture-coupled 15, 184, 185, 188, 190,
    246, 263
AR bandwidth 120, 123, 130, 132, 135,
    136, 137, 138, 140, 156, 157, 158,
    160, 219, 234, 235, 236, 237
asymmetric bent dipole 70, 73, 74, 76, 79
axial ratio (AR) 120, 131, 133, 156, 206,
    219, 278, 281, 283, 284

balanced 7, 8, 9, 12, 24, 66, 125, 127, 129,
    130, 131, 132, 133, 134, 135, 138,
    142, 146, 147, 148, 149, 150, 151,
    152, 153, 154, 155, 264, 271, 272,
    280, 283
balanced circularly polarized loop
    antenna 7, 125
balun 19, 65, 66, 68, 70, 73, 75, 127, 129,
    130, 132, 134, 146, 147, 150, 154,
    155, 176, 235, 264, 266, 281, 282
band-notched antennas 7
base station antennas 8, 110, 145, 149,
    222, 239
Bessel-Fourier series 17, 43, 218
bi-conical antennas 21, 22, 43, 68
bidirectional, bisensing circularly
    polarized (CP) radiation 12
binary phase shift keying (BPSK) 221
bit error rate (BER) 221
boundary conditions 2, 13, 15, 18, 29, 31,
    34, 92, 93, 105, 116, 217, 283
broadband antennas 21
broadcasting 8, 13, 28, 79, 143, 238, 240,
    248, 276

broadside coupled stripline (BCS)
    147, 150, 164, 272
bulbous antennas 21

capacitive probe compensation 184
car-to-car (C2C) 240, 265, 275, 281
cavity model 184, 185
centro-symmetric 13
circular loop antennas 13, 111, 112, 270,
    278, 280
circularly polarized (CP) antennas 4, 17,
    155, 219, 263, 268, 270, 272, 273,
    278, 282, 283
circularly polarized loop-dipole
    antennas 144
circularly polarized microstrip patch
    antennas 263, 284
circular sector patch antennas 283
code division multiple access 2000
    (cdma 2000) 149
co-linear dipole array 183
common E-mode (CE-mode) 4, 5, 7, 8, 11,
    12, 13, 16, 31
common M-mode (CM-mode) 6, 13, 15,
    16, 17, 18, 24
compact patch antennas 281
complementary dipole antennas 7, 11, 12,
    143, 144, 145, 146, 147, 149, 151,
    153, 155, 157, 159, 160, 161, 163,
    165, 167, 169, 170, 171, 173, 174,
    175, 176, 177, 179, 181, 241
computer-aided design (CAD) 185
constant aperture antenna 68
constant gain antenna 76
consumer electronics devices 231
continuous spectrum 22

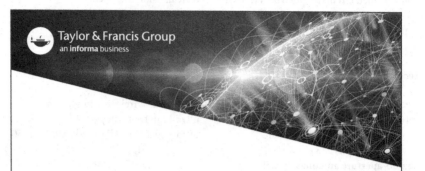

Printed in the United States
by Baker & Taylor Publisher Services